Issiakou Moussa

Improving farming methods

Issiakou Moussa

Improving farming methods

adapted to tropical and subtropical regions

ScienciaScripts

Imprint
Any brand names and product names mentioned in this book are subject to trademark, brand or patent protection and are trademarks or registered trademarks of their respective holders. The use of brand names, product names, common names, trade names, product descriptions etc. even without a particular marking in this work is in no way to be construed to mean that such names may be regarded as unrestricted in respect of trademark and brand protection legislation and could thus be used by anyone.

Cover image: www.ingimage.com

This book is a translation from the original published under ISBN 978-613-8-43449-8.

Publisher:
Sciencia Scripts
is a trademark of
Dodo Books Indian Ocean Ltd. and OmniScriptum S.R.L publishing group

120 High Road, East Finchley, London, N2 9ED, United Kingdom
Str. Armeneasca 28/1, office 1, Chisinau MD-2012, Republic of Moldova, Europe
Printed at: see last page
ISBN: 978-620-6-19756-0

Copyright © Issiakou Moussa
Copyright © 2023 Dodo Books Indian Ocean Ltd. and OmniScriptum S.R.L publishing group

CONTENTS

SUBJECT OF THE SUMMARY ASSIGNMENT

Theme:

Most countries in the tropics and subtropics have agricultural economies.

The second half of the 20th^{ème} century was a time of determination and the quest for independence. The study of the evolution of peoples shows that this political or economic independence upsets ideas and jeopardizes the extensive economy.

A degree of stability can only be achieved through an "agricultural revolution" that puts "the natives" in a position to improve their means of production, profitability and marketing.

This improvement requires the development and application of research methods and cultivation techniques often influenced by the theories and results of temperate countries.

If, in nature, nothing is created, nothing is lost, everything simply transforms itself, then any intensive farming technique presupposes adaptations to local conditions (soil, climate, equipment, people's educational evolution).

As a sub-engineer, you'll be studying and proposing operating methods adapted to these regions.

The aim of this assignment is to enable us to assess your ability to adapt to the demands of these regions.

<u>Theme summary:</u>

We summarize the theme as follows:

<u>Contribution to the improvement of farming methods adapted to tropical and subtropical regions.</u>

All the work in this document is based on this summary.

Work to be done

1)   What are the essential pedological characteristics of soils in tropical regions, or

of a more specific region?

How can we improve the agricultural value of these soils?

2) How can biology and genetics be considered as a tool for agricultural production?

3) Mechanization is an important factor in the profitability of agricultural production and a factor in the social evolution of workers. Are you in favor of mechanization in these regions?

Where? When? And how?

4) Soil fertilization and plant protection are key factors in production. What are the principles governing their use?

Using examples, describe the results obtained by using these products (fertilizers and herbicides).

5) What are the basics of running a profitable farm in the tropics?

6) Workforce hygiene and well-being are influential factors in productivity. How should you view them in tropical regions?

7) Choice between points 1 and 2.

1. For a choice of food crops and industrial crops.

a) Draw up a labour schedule for these 2 crops (from sowing to harvesting, including any harvest treatment);

b) Factors likely to improve these crops ;

c) Economic opportunities for better marketing.

2. What livestock can be successfully raised and farmed in the region (breed, breeding method, products obtained)?

What are the main cattle diseases?

Food is a production factor. How can we feed these livestock and improve their nutrition?

Planning the breeding and exploitation of this livestock.

Possibility of income pricing.

THANKS

Throughout my training at EDUCATEL INTERNATIONAL, I have enjoyed a great privilege. This training could not have been completed without the help and sustained actions of the Eternal GOD and certain people to whom I remain infinitely grateful.

It therefore gives me great pleasure to express my sincere thanks:

- to the Director General of Educatel for accepting me into his school;

- Mrs Roslyne BOIVIN, my advisor, for her unfailing advice and support;

- to all Educatel and CODIFOR staff for their efforts;

- To my various teachers, who spared no effort to ensure that we completed our training, I would like to express my gratitude;

- to the members of the jury, who, despite their busy schedules, agreed to assess my Summary Work;

- to all the staff of my department (DEDRAS-ONG) for their moral support;

- to my friends: Jean KPETERE, Alain TROUKOU, Patient COUBEOU, Benjamin BOUSSO, Augustin NASSAM for their commendable contribution to this study,

- my wife Hélène and my daughter Gariya, who encouraged and supported us,

I'd be remiss if I didn't mention the farmers, agricultural extensionists and other people I met who took the time to answer my various questions.

Finally, to all those who have contributed in any way to my training, please accept my sincere thanks.

FOREWORD

This document marks the end of my training at Educatel International. It constitutes the Summary Work at the end of my study as a **Sub-Engineer in Tropical Agronomy**.

Its completion has enabled me to deepen my knowledge of farming methods adapted to **tropical and subtropical** countries.

At the end of this training course, I can't claim to have produced a flawless summary, free of imperfections. It is for this reason that I ask the indulgence of the professionals and readers of this document in the formulation of their criticisms and suggestions in order to enrich it.

INTRODUCTION

This document, which represents the Summary Work at the end of my training, is in two parts: **the literature review and the field results/discussions**. It deals with the problems hindering the proper development of agriculture and proposes solutions for improving farming methods adapted to countries in tropical and subtropical regions, in order to promote the sustainability of this agriculture.

The theme proposed for this summary work is highly relevant. Indeed, the relevance of this theme is justified in the sense that it has been noted that agriculture is the main source of subsistence products and economy for most countries in tropical and subtropical regions. However, this observation raises the following questions:

- do current agricultural practices contribute to soil conservation and environmental protection for **sustainable agriculture** in tropical regions?

-Are there   other, more suitable and economical practices?

- Are production factors available and favorable?

- are crop yields good?

- Is the market for your products favorable?

- does agriculture really feed people?

In developing our theme, we shall attempt to provide answers to all these questions. Through this summary work, we hope to contribute to the improvement of farming methods adapted to tropical and subtropical regions.

To do this, we will first review existing operating methods, then, on the basis of a rational analysis, we will make proposals for adapting these methods.

Issues

In the past, as today, the economies of certain tropical and subtropical countries are essentially based on agriculture. No viable development policy for these countries is possible if the agricultural sector is neglected. In Benin, for example, it accounts for 45% of Gross National Product (GNP), over 90% of export earnings and provides employment for around 75% of the working population.

According to ICRA TOGO (1990)[1] , Togo's GNP was estimated at 340 million CFA francs in 1986, 32% of which came from the agricultural sector, which employs 80% of the working population. Agriculture is therefore the foundation of the economy of certain tropical and subtropical countries, and the driving force behind their growth. In fact, the United Nations Development Programme stated in its May 1989 identification mission report that there is currently no alternative to agriculture in the process of economic growth.

In the tropics, the most diverse types of agricultural production coexist. The most widespread is the small family farm, growing mixed food crops mainly for self-consumption. At the other end of the scale are large-scale salaried plantations specializing in monoculture for export. In the intermediate stages, these two forms of production are combined. Similarly, technical levels are extremely varied. It's true, there are many different production systems in the tropics, as pointed out by the last world census of agriculture carried out in 1970 under the aegis of the FAO; but tropical agriculture has enormous constraints.

Climate: the peculiarities of tropical climates give them their own specific characteristics. The regularly high temperature in most of this zone plays a negligible role in the distribution of production.

The absence of very low temperatures encourages intense parasitism, especially animal parasitism, which attacks crops, domestic animals and humans (malaria, fever, bilharzia...).

Rainfall is an essential factor in crop distribution. But in the tropics, often abundant rainfall combined with high temperatures leads to rapid decomposition of organic fertilizers and significant leaching of soluble fertilizers. Heavy rainfall, especially at the start of the season, is a major factor in erosion.

Soil: in tropical regions, rainfall is often brutal and abundant, causing continuous

---

1 ICRA TOGO (1990), Contribution à l'étude des stratégies de production dans le système agricole de la zone d'AGOU au TOGO.

elimination (leaching) of chemical elements essential to plants.

Poor farming practices: failure to respect fallow periods due to lack of arable land, low use of improved seeds, failure to adopt popularized technical innovations, difficult access to specific fertilizers and low use of organic fertilizers make soils poor and reduce production.

Labor: labor is an important factor in agricultural production. It is becoming increasingly scarce, not only because of the rural exodus, but also because of the precarious state of health of the able-bodied.

(Monetary) capital: Without agricultural credit, many farmers are unable to acquire the means of production (mechanization, for example) with their monetary resources alone.

As a result, traditional agriculture continues to grow, despite its low yield, because it is less mechanized or not mechanized at all due to a lack of financial resources.

Marketing: marketing has always been and will remain the driving force behind agricultural development. But marketing channels are only organized for export products. Many tropical countries have difficulty selling a modest part of their agricultural production on highly competitive foreign markets, while importing food products for their urban consumers, or even to avoid famine in an area hit by climatic hazards.

Agricultural research: For several decades, agricultural research stations have been developing "adapted" cultivation techniques based on theories and results from temperate regions.

All the constraints mentioned above are not conducive to good farming practices adapted to tropical and subtropical regions.

As a result, despite being the mainstay of tropical economies, agriculture remains extensive and uncompetitive.

The summary work I propose to do will help identify the main technical and economic problems facing farmers in tropical and subtropical regions, and make proposals for solutions. In particular, this work aims to clarify the difficulties farmers face in satisfying their respective interests with the resources at their disposal.

To do this, we'll be focusing on how farmers combine a number of agricultural activities and techniques on their farms, given the diversity of edaphic conditions and more or less predictable variations in climate. A farmer's profession requires the ability to perform a large number of very different tasks (tillage, fertilization, crop protection, animal care, equipment maintenance, crop preservation, product processing and marketing, etc.), all of which need to be well organized in terms of space and time. Not only do farmers need to acquire a multitude of specific skills concerning the various plant and animal productions, but they also need to be able to combine their multiple activities in the best possible way, while retaining control of the whole.

It would therefore be wrong to claim to be able to design and formulate interventions appropriate to the conditions and interests of farmers in tropical regions without a good understanding of the complexity and performance of their farming methods. This is why we need to know in advance how the various available resources (soil, climate, labor force, capital...) are allocated to the different activities, evaluate the economic results and make proposals to improve farming methods adapted to tropical regions.

Working hypothesis

There are farming methods adapted to tropical and subtropical regions.

These methods remain theoretical and not very operational.

Current production conditions in tropical and subtropical regions are not conducive to the proper application of adapted farming methods.

Objectives

The purpose of this summary is to :

1- Analyze farming methods used in tropical and subtropical regions

2- Propose farming methods adapted to the realities of tropical and subtropical regions.

Work methodology

The methodological approach adopted for the production of the summary work at the end of my training is based on the following five (5) steps:

- Reading and respecting the general end-of-studies regulations ;

- Documentation ;

- Field research ;

- Data analysis ;

- Suggestions for improvement methods.

Reading of the general end-of-study regulations

As soon as we received the summary assignment document, we carefully read through the general end-of-study regulations, the theme and the assignment.

This made it possible to :

- to better understand and summarize the theme;

- take into account the provisions stipulated in the general regulations;

- draw up a substantial work program;

- select a plan for presenting the summary work.

Documentation

Firstly, the sub-engineering courses containing the points of the work to be carried out were exploited. These included :

- Tropical Agriculture Volume 1

This book provides information on :

- Soil classification in tropical regions,

- Chemical and biological characteristics of tropical soils,

- Principles of use of fertilizers and plant protection products,

- Soil science, soil properties

In this work, we have identified :

- Physical, chemical and biological properties of soils,

- Soil fertilization in tropical regions,

- Tropical Agriculture Economy

The passages in the document that caught our attention concern the marketing of tropical products.

- Fertilization and Fertilizer

This document has enabled us to deepen our knowledge of the principles of tropical soil fertilization.

- Agricultural machinery Tome1

In this work, we have identified :

- Different types of soil preparation,

- Farming tools.

We then turned to documents dealing with agriculture in the tropics. We will mention here three of the works most frequently used in our work.

- Le Mémento de l'agronome editions 96

- Soils and their agricultural potential (INRAB, 1995)

- Hygiène pour nos plantes, aussi prévenir vaut mieux que guérir (GRAD, carnets écologiques N°10).

The documentation enabled us to :

- deepen our knowledge on the theme of summary work;

- identify an issue on which the summary work is based;

- complete the results obtained from reading the courses;

\-   identify the information to be sought in the field.

On the whole, the data used in the documents concern the tropical region. This is defined by warm climates, located on either side of the equator between the two tropics (23°27' North and South), in a more or less regular band that circles the globe. There is general agreement on the boundaries of this region, where the average temperature in the coolest month does not fall below 17 to 18°C, or 20°C.

Field research

As part of the summary work, a number of agricultural stakeholders living in the department of Borgou/Alibori (northern Benin) were surveyed. These were :

\-   practicing farmers, i.e. peasant farmers;

\-   agricultural extension workers ;

\-   engineers.

In order to ensure that field surveys are properly conducted and data collected, a form has been designed with a series of open-ended questions (form in appendix).

The information collected relates to :

*** among farmers**

\-   Main crops and production targets

\-   Crop soil typology based on endogenous knowledge

\-   Local farming practices

\-   Endogenous food preservation and storage techniques

\-   Marketing channels

\-   Farm business management

Supporting the development of agricultural production

*** Among extension workers and engineers**

\-   The nature and role of support for producers

- Intervention approaches

- The main results and

- The difficulties of supporting producers

■ Survey area

The Borgou/Alibori department, where the survey took place, is the largest in Benin. It borders Nigeria to the east, Atacora/Donga to the west, Collines to the south and Niger to the north. It has a total population of 1,046,645, of whom 798,241 (76%) live in rural areas. The department's main activity is agriculture, and the main crops grown are yams, corn, cotton, sorghum, cassava and fruit trees such as mango and cashew.

The farmers surveyed are all from this department and their main activity is farming.

The extension workers interviewed were Agents Polyvalents de Vulgarisation (APV) working in the Borgou/Alibori department, and more specifically in the communes of Tchaourou, Banikoara, Nikki and Parakou.

The agronomists referred to here are those with at least five (5) years' field experience and currently working in the department where the survey took place.

Data analysis

This is a comparative qualitative analysis. The results obtained during the survey were compared with those found in the literature before drawing a conclusion and making proposals.

Suggestions for improvement methods

After having familiarized ourselves with the harvesting methods used in tropical regions through documentation and field research, and their advantages and disadvantages, we have made proposals with the aim of contributing to the improvement of these methods.

Part I: DOCUMENTARY REVIEW

I - ESSENTIAL PEDOLOGICAL CHARACTERISTICS OF SOILS IN TROPICAL REGIONS

* What is a floor?

According to the Petit Larousse, soil is the superficial, loose part of the earth's crust, resulting from the transformation, in contact with the atmosphere, of the underlying layer (parent rock), and subject to erosion and human action.

It is the essential basis of all agricultural production, providing plants with support and a reserve of water and nutrients.

Soil covers a large part of the world's continents. In terms of human needs and health, it performs four (4) groups of essential functions (source: CIRAD and GRET, 96. Mémento de l'agronome):

-   **Biological functions:** the soil is partially or completely home to numerous animal and plant species, and many biological cycles include the soil. Soil's biological activity is essential to its construction, functioning and fertility (aggregation, porosity, availability of nutrients, etc.). Soil cannot exist without abundant and diversified biological activity.

-   **Food functions**: the soil produces and contains all the elements[2] of life; it accumulates and then makes available to plants and animals most of these elements. Much of what plants need for life is contained in the soil. Plants use the soil at depths of several meters.

-   **Exchange and filter functions**: soil is a porous medium through which water and gas flows constantly. Water from wells, pumps and springs has already passed through the soil, and soil porosity influences its supply. In addition, soil is a filter; as water passes through it, it is transformed, and the chemical and biological quality of the water remains unchanged.

---

2 Calcium, potassium, iron, nitrogen, carbon dioxide, water, air, etc.

depends on soil properties. Soil is also in constant exchange with the atmosphere.

-   **Material and support functions:** soil is often a building material: sand, clay, ferruginous armour, crusts. It is both a support and a material for buildings, roads, dams, canals and pottery.

In short, for human societies, soil is a source of life. Through its plants, animals, water and minerals, the soil nourishes people, and their health and activities depend on it. It is therefore essential to[3] know it and know how to use it.

Tropical soils come in a variety of classes.

1-1 Soil classification in tropical regions

According to the French classification[4] , tropical soils comprise twelve classes:

- Raw mineral soils,

- Poorly developed soils,

- Vertisols,

- Andosols,

- Calcimagnesic soils,

- Isohumic soils,

- Burnished soils,

- Podzolized soils,

- Soils with iron sesquioxides,

- Ferralitic soils,

- Hydromorphic soils,

---

3 Tropical soil which is the subject of this work.
4 Established by French soil scientists in Madagascar and French-speaking Africa in collaboration with soil scientists working in France.

- Sodic or halomorphic soils.

Note that the classes group together soils from the same formation process.

According to the same classification, the surface distribution of tropical soils in the various soil units highlights the importance, in descending order, of the following soils:

- **Crude mineral soils**: the profile of these soils is of the A/C type[5] , mineral matter is more or less disaggregated, organic matter is rare. These are desert soils, soils on rock that have not or cannot undergo pedological evolution. They are of little interest to agriculture.

- **Poorly developed soils**: the profile of these soils is of type A/C. It shows a clear humus-bearing horizon, fairly extensive alteration of the parent rock, but an evolution that is difficult to define due to the youth of the soil or the lack of action of the climate. These soils are found in regions where erosion is strong, or in places where river deposits are frequent. These soils can be very interesting for agriculture.

- **Ferralitic soils**: the profile of these soils is of the A/B/C type[6] , often very thick. They develop in the wettest regions of the tropics. Minerals are highly altered, with the release of iron, manganese and even aluminum, and the start of silica. The PH is acidic. These soils are planted with oil palms, rubber and cocoa trees.

- **Iron sesquioxide soils**: the profile of these soils is of the A/B/C type. These soils are characterized by the individualization of iron, which gives them a striking red or ochre color. Organic matter content is low. They are divided into Mediterranean red and brown soils and tropical ferruginous soils. These soils are often used for food crops (corn, yams, etc.).

- **Brunified soils:** in tropical zones, only brown, entrophic soils derived from calcium-rich rocks are found. The soft humus is well bound to the mineral matter, but the color tends to reddish-brown due to the release of iron sesquioxides. These soils

---

5 Soils with few differences: The A horizon contains organic matter and rests directly on the bedrock, i.e. the C horizon.
6 Advanced soils: the B horizon is enriched with fine or soluble elements leached from the A horizon.

are of good fertility.

- **Vertisols**: these are dark-colored soils with an A/B/C profile. They are characterized by the presence of swelling clay (tropical black clay). In the dry season, they show wide shrinkage cracks up to 50cm deep, the lower their permeability. They are difficult to work. These soils are well suited to cotton cultivation.

- **Ferry soils**: intermediate between ferralitic and ferruginous soils. They show an accumulation of clay.

It should be noted that ferralitic and ferruginous soils are those frequently found in the tropics.

* Case study of Benin and Togo soils[7]

Soils of Benin

In Benin, there are five (5) dominant soil categories, whose genesis and evolution are the result of the combined action of a number of factors such as climate, plant formations, parent rocks, landforms and geomorphological history.

These different soil categories are divided into agro-ecological zones.

***Littoral and fluvio-lacustrine sandy zone**: this zone comprises the coastal sandy strip and the fluvial and lacustrine alluvium of the Mono, Ouémé and Atlantique departments.

Coastal sandy soils are poor in organic matter and have low exchange capacity and water retention capacity. They can only be valorized by adapted perennial species such as coconut, filao, eucalyptus, acacia, etc.

Hydromorphic soils are found in this zone. They have light, clayey textures and good chemical properties. They are well suited to market gardening, corn and rice cultivation.

***SudanoGuinean zone on Vertisols**: this corresponds to the Lama swelling clay depression. This type of clay gives the soil high cation exchange capacity and good

---

7 From INRAB, 95 and ICRA Togo, 90

structural stability. These soils have good chemical properties and are well suited to maize, cowpea, sugarcane and market garden crops.

***SudanoGuinean zone on bar soil**: this is the plateau area (southern Benin). The soils of this zone are known as ferralitic soils or terre de barre. These are sandy-clay soils with a stable structure and low water retention capacity. Their fertility potential is average. When heavily exploited, they degrade and become unproductive, requiring regeneration. Crops such as oil palm, coffee, maize, cassava, groundnuts, cowpeas and forest species such as teak respond well to these soils.

The use of these soils requires periodic applications of potassium-dominant mineral fertilizers, organic fertilizers or the return of crop residues to the soil.

***Transitional Sudano-Guinean zone**: entirely occupied by leached or impoverished tropical ferruginous soils.

The main crops grown on these soils are maize, manioc, yams, groundnuts, cowpeas, cotton and forest species such as cashew.

***Sudanian zone of northern Benin: the** largest in the country. According to the pedological division, this zone includes :

- *Northeastern Sudanian zone:* in this zone, soils are mainly tropical ferruginous soils with highly variable agronomic characteristics. These are fine clay and sand soils.

The average usable depth may be limited by the discontinuity of a sometimes massive concretion horizon, making for poor chaining, which is generally average.

These soils have better chemical fertility, but their physical properties are often too restrictive for plants.

Ferralitic and hydromorphic soils are also found in this zone.

The main crops in this zone are sorghum, maize, yams, groundnuts, cowpeas, cotton, shea, cowpea and cashew.

- *Northern Sudan zone:* tropical ferruginous soils are also found in this zone.

Where useful depth and texture allow, these soils are home to fine plantations of fruit trees and forest stands.

-    Northwest Sudanian zone: Here, crude mineral soils and tropical ferruginous soils make up the bulk of the soil cover. Crops grown in this zone include maize, sorghum, millet, groundnuts, cotton, etc.

-    *Extreme northern Sudano-Sahelian zone:* this zone is characterized by poorly developed soils, hydromorphic clay soils with high hydro-agricultural potential, and tropical ferruginous soils. Potatoes, ogiois, rice, millet and sorghum are cultivated in this zone.

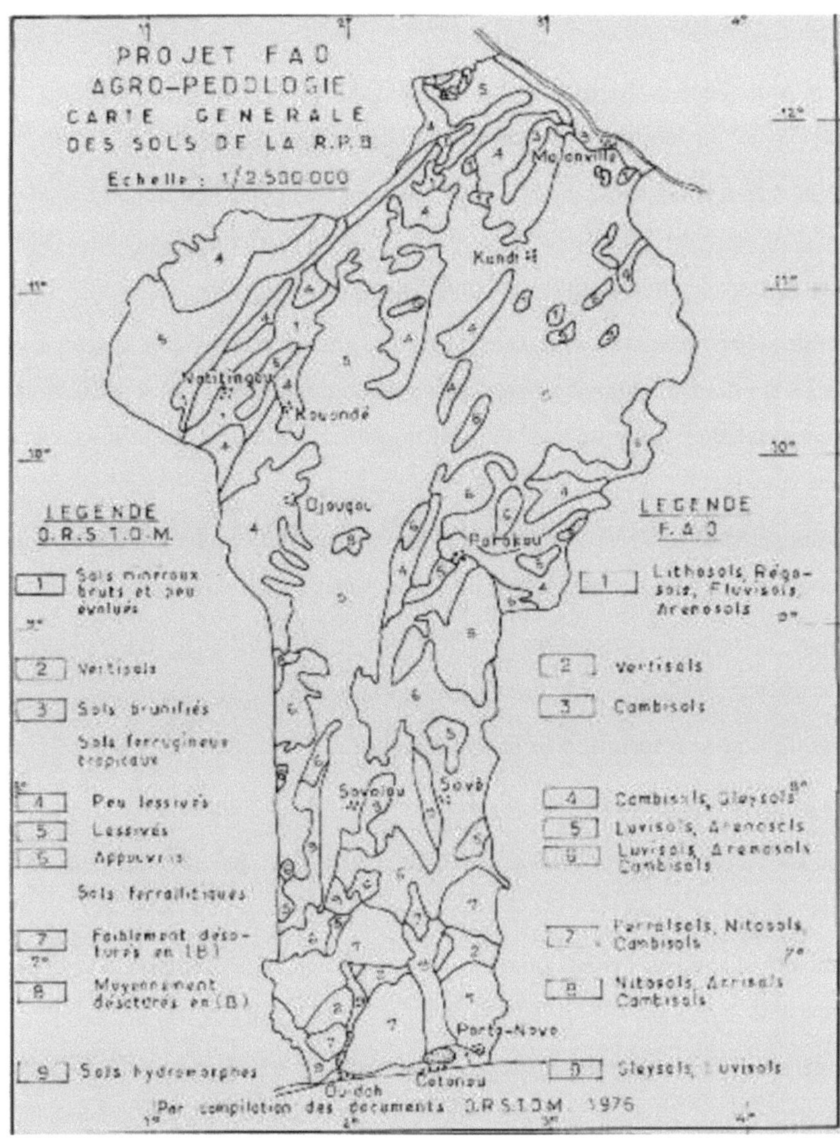

General soil map of Benin

Soils of Togo

Togo is characterized by a great diversity of soil types. This diversity is closely related to the variability of the natural environmental factors that govern the evolution of these soils, essentially climate, parent rocks, landforms, vegetation and

23

human activity.

The main soils found in Togo, according to the pedological classification, belong to eight (8) classes, the most important of which in the field of agriculture are :

***Tropical ferruginous soils:** these are the most abundant, but not the most fertile; very fragile, they lend themselves to food crops in traditional systems. Their disadvantages are declining fertility and rapid lateralization.

***Ferralitic soils**: These are found mainly in the south and southwest of the country. These soils are deep, have good physical properties and some of the most satisfactory hydric characteristics. They are ideal for growing coffee, cocoa, oil palm, citrus, corn, yams and cassava.

***Hydromorphic soils**: These cover a small area, but are found in the south and north of the country. They have a high potential for rice, sugar cane and market gardening.

Soil cover in tropical regions has essentially physical, chemical and biological characteristics.

1-2 Physical characteristics of tropical soils

Texture and structure determine the physical characteristics of tropical soils.

Tropical soils are made up of solid elements (clay, silt, sand, limestone, humus) and fluids (air, water).

The structure of soil cover exists from the scale of particle organization to that of the landscape unit.

The structure of tropical soils is fragile and poorly developed in semi-arid zones, much more resistant thanks to iron and organic matter in humid zones. In general, the types of structure most widespread in tropical regions are quite favorable to agriculture, notably fragmentary structures whose horizon well accommodates the root system, since the aggregate structure is rounder, finer and less consistent[8] . This structure facilitates the circulation of gases and solutions, and the penetration of roots

---

8 This commentary is taken from Memento de l'agronome, 96

and animals.

It is favoured by :

- The presence of neutral or slightly basic organic matter;

- An absorbent complex saturated with calcium ;

- High biological activity.

This structure generally provides the best conditions for crop development, since it is stable.

Structural stability seems to be an essential characteristic of the physical state of tropical soils. But human action (cultivation, bush fires and deforestation) and natural phenomena (erosion) are reducing this stability.

Structural instability is a danger: it hinders plant rooting, animal biological activity and fluid circulation, particularly the vertical penetration of water into the soil. In short, the soil's food resources, including water, are only regularly accessible to plants if the structure is stable.

The texture of tropical soils is generally not, on its own, a determining factor in crop success. But the combination of properties induced by structure and texture can be decisive, particularly for water penetration and retention, and root establishment.

Tropical soils are generally 20-30cm deep or more.

Generally speaking, the physical characteristics of soils in tropical regions are good in humid zones and sometimes very good, as in the case of ferralitic soils. They are less so in dry zones due to the fragility of structures (little organic matter), the abundance of sandy soils and the presence of very clayey soils (vertisols, brown soils, entrophes), which are difficult to cultivate.

In addition, the profile of tropical soils corresponds to the following diagram:

<u>Diagram 1</u>: Profile of a tropical soil

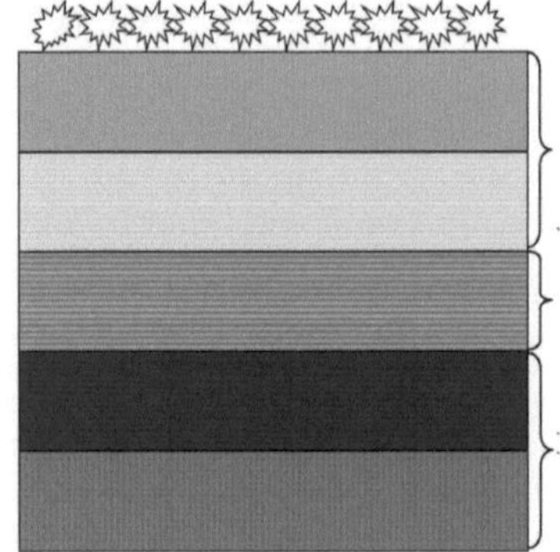

Horizon A: The surface horizon that is leached and depleted, but also continually enriched by the decomposition of dead plant and animal debris,

Horizon B: the accumulation horizon. It is enriched with substances from horizon A that have been washed down by the leaching process.

Horizon C: original material resulting from alteration of the parent rock

- **Proposals for improving tropical soil stability**

According to the authors studied above, for tropical soils to be stable and favourable for agriculture, the following must be avoided:

- Cropping systems that require fields to be worked in conditions that are too dry (risk of spraying) or too wet (risk of compaction) - Excessive irrigation followed by excessive drying out.

We must promote :

- Alternating crops allow alternating types of rooting

- Tillage that aerates without compacting

1-3 Chemical and biological characteristics of tropical soils

Tropical soils have low levels of the biogenic elements nitrogen, phosphorus, potassium, calcium, magnesium, sulfur and trace elements (copper, zinc, molybdenum). They also have a low retention capacity for mineral cations.

It should be noted that the stock of assimilable elements in the soil is low in tropical conditions: 3 milliequivalents of exchangeable bases per 100 grams of soil represent an acceptable average value in ferralitic soils, whereas a dozen milliequivalents characterize exceptional fertility in ferralitic environments. At the same time, 40 to 80 milliequivalents or more can be found in certain temperate enriched soils (e.g.

market garden soils).

Tropical soils are poor in chemical elements due to :

- rainfall that is often brutal and abundant, resulting in a chain reaction that continuously eliminates the chemical elements essential to plants

- the presence of kaolinite as the dominant clay. In fact, this clay has a low exchange capacity of 5 to 10 milliequivalents per 100 grams of clay, and can therefore only retain cations in small quantities.

- the retrogradation of insolubilization removes a significant amount of phosphorus from the soil and from the fertilizer to the cultivated plants.

Soils in tropical regions are much less well supplied with exchangeable bases. Cultivation causes a significant reduction in exchangeable bases (Ca, Mg, K).

This decline has four main causes[9] :

- Crop removal: especially important for potassium

- Mineral fertilizers, especially nitrogen fertilizers (this affects Ca and Mg in particular). They are mainly used on cash crops (cotton, peanuts, coffee, rubber, oil palm, cocoa, etc.).

- Erosion by stripping the richer surface layer of soil.

- Soil exchange complex is reduced, and base leaching does not follow. The decrease in organic matter content under cultivation can often affect up to half of the fixation sites.

The reduction in exchangeable bases in the soil obviously causes a drop in PH, all the more so as the original PH is generally quite low (usually between 4.5 and 6) and the buffering capacity of soils is low.

The plant and animal debris that falls on the soil is the essential source of organic matter, which is gradually transformed by mineralization into soluble or gaseous elements that can be assimilated, and by humification into humic colloidal

---

9 According to Tropical Agriculture Tome 1 (EDUCATEL)

complexes.

These are slowly mineralized to feed the plants on the soil and to block the biological cycle of nitrogen and carbon, which can be more or less rapid, discontinuous or not.

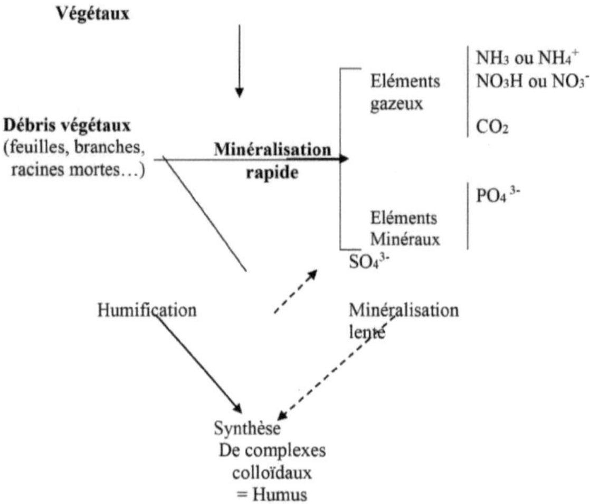

Diagram 2. Evolution of fresh organic matter in the soil (Source: Pédologie : propriétés des sols, course de sous ingénieurs en agronomie tropicale)

As mentioned above, human action, natural phenomena, the decrease in organic matter content and the loss of bases followed by soil acidification, are the main causes of soil fertility decline in the tropics.

Appropriate solutions for improving the agricultural value of tropical soils are needed to overcome these drawbacks.

1-4 Potential for improving the agricultural value of tropical soils

According to the literature consulted, the following proposals have been made to improve the agricultural value of tropical soils.

*   *Fallowing:*

Fallow land means leaving the soil to rest for several years to allow it to regain its natural fertility.

The system of fallowing, essential to ensure a minimum subsistence, is still important for people living in the tropics.

Fallowing provides good soil cover and protects the soil from all forms of erosion (rain and wind). After a few years, it gives the soil its natural fertility, reducing the need for chemical fertilizers.

*   *Crop rotation.*

Rotation is the order in which crops are grown on the same soil over a certain number of years, known as the rotation cycle.

It should be noted that continuous cropping has two main effects on the soil: a rapid reduction in organic matter and a drop in exchangeable bases, resulting in soil acidification.

*   *Setting up perennial and annual crops*

Soil conservation and fertility are ensured by good soil cover with plants.

Cereal crops (maize, sorghum, millet) and legumes (groundnuts, cowpeas) are annual crops. During their vegetative cycle, these crops protect the soil against erosion. After harvesting, the residues (leaves, stems, roots) can be left in the field, where they rot under the effect of humidity and the action of micro-organisms, giving the soil humic and chemical fertility.

*   *No overgrazing or bush fires.*

Overgrazing degrades soil structure and destroys plant cover. As a result, the bare soil is subjected to the erosive action of water.

In tropical regions, bush fires are a frequent occurrence. This fire not only deprives animals of their food and shelter, but also burns organic matter and kills the micro-organisms responsible for its decomposition. The soil becomes bare and dry. As a result, it is exposed to erosion. It also becomes poor and unprofitable for agriculture.

From the above, we can see that tropical soils have both favorable and unfavorable characteristics for agriculture. Consequently, the improvement of these soils for their agricultural value is necessary. But is improving the agricultural value of tropical soils enough to make tropical agriculture profitable? One of the preoccupations of the structures supporting agricultural development is the search for varieties that meet the needs of farmers, consumers, industries, traders, processors and political decision-makers. Hence the need for biology and genetics.

II- <u>BIOLOGY AND GENETICS: A TOOL FOR AGRICULTURAL PRODUCTION.</u>

According to the Petit Larousse VEUF 2001, **biology** is the science of life and specifically of the reproductive cycle of living species. **Genetics is the** science of heredity, based on the theory of genes. From the above, genetics is one of the most complex areas of modern biology. In practice, it leads to a set of processes which, from a group of individuals lacking certain sought-after traits, enable us to obtain another group of individuals, more or less productive: the variety - thus bringing about progress. It thus enables plant improvement.

Plant improvement began with agriculture and progressed with biology.

Schematically, Evans (1976)[10] distinguishes three stages in the history of the process of improving cultivated plant species:

• domestication takes place in the environment of origin; the wild plant becomes suitable for cultivation if it is capable of expressing a small number of characteristics grouped together under the term domestication syndrome

• the spread of this plant outside its domestication center requires sufficient genetic plasticity to adapt successfully to new growing environments

• finally, genetic improvement accompanies the evolution, generally intensification, of cropping systems, which creates a demand for genotypes with superior productive potential.

Generally speaking, the introduction of improved seeds is an innovation. According

---

10 Henri Hocdé, Jacques Lançon, Gilles Trouche (2001), Actes de l'atelier sur la Sélection Participative : Impliquer les Utilisateurs dans l'amélioration des plantes (Proceedings of the workshop on Participatory Selection: Involving Users in Plant Improvement).

to Barbara Bentz *et al.* (2002)[11] , certain conditions are decisive for the transition from a simple technical "novelty" to an innovation that spreads. The latter must :

- make better use of the farm's resources,

- increase income or labor productivity,

- reduce risk and secure the family economy (consumption, cash flow) at key moments.

Innovation is not always about simply increasing yield or production (intensification is not an end in itself). Above all, it must enable farmers to better achieve their objectives, under conditions that are not too risky for them.

Lastly, innovation only occurs if the novelty is viable in the socio-economic, political and cultural context, etc. Quite often, it's not purely technical aspects that prevent innovation from taking place, but constraints of another kind.

In short, biology and genetics are tools for improving plant production, both quantitatively and qualitatively. These instruments are used in agricultural research stations for the selection of improved and resistant seeds, whose appropriation by users is low.

Given the high rate of population growth, the shortage of arable land and low crop yields in some tropical countries, if agricultural research is not given a high profile in these countries, food insecurity will prevail and development will be jeopardized.

It's true that biology and genetics are seen as a tool for agricultural production. They promote agricultural profitability. But alongside these, factors such as farm mechanization must be used to make agricultural production in tropical regions more profitable.

III- <u>MECHANIZATION OF AGRICULTURE IN TROPICAL REGIONS</u>

---

11 Barbara Bentz et al (2002), Supporting Farmer Innovation, Guide pratique Agridoc

3-1 Mechanization: an important factor in the profitability of agricultural production

In tropical regions, agriculture has long remained traditional. Man uses rudimentary tools (hoe, cutter, etc.) to carry out all agricultural activities. This practice is not only tiring and slow, but also fails to produce the best yields.

G. Le Thiec (1996)[12] states that over 70% of Third World farmers currently use only hand tools. Although these can be improved, they do not greatly enhance work performance.

The introduction of mechanization has gone some way to correcting the shortcomings of hand tools. The mechanization of agriculture relies on animal power and "motorization".

- Animal energy

The use of draught animals requires greater power than hand tools (work and speed). It facilitates transport, an essential activity both on and off the farm. It encourages the sedentarization of animals on the farm, leading to changes in production systems. It also provides products for human consumption (milk and meat) and soil fertilization (manure).

Cattle (especially males), horses, donkeys and dromedaries are the main animal species used for traction in tropical agriculture. The draught animal species selected is the one that is available locally, well known to potential users, hardy and adapted to the pathology of the area, or the one that offers the best possibilities of adaptation if it is necessary to introduce it.

The main items of equipment used in animal traction are: plough, harrow, seeder, Canadian, cart (for picking up crops).

- Motorization

In industrialized countries, draught animals have gradually been replaced by engines, eliminating the less efficient players from agricultural production.

---

12 G. Le Thiec (1996). Agriculture africaine et traction animale, found in Mémento de l'agronome

Agricultural motorization encompasses everything to do with the use of engines to carry out agricultural work[13] :

- Tractors with their equipment and self-propelled machines;

- Tillers and other specialized machinery;

- Motors to drive machines used in stationary, back-pack operation. In developing countries, the acquisition of complete motorized chains is often beyond the technical and financial reach of farmers. On a national scale, agricultural motorization poses problems that go far beyond the technical choice and cost of equipment. It's a question of choosing the most appropriate formulas for achieving economic development objectives. We need to take into account the environment in which this motorization will be used, assess the benefits of its introduction, study the conditions for its implementation and propose the appropriate power levels and equipment.

On the whole, agricultural mechanization mainly concerns soil cultivation. However, there are agricultural machines used for harvesting crops, which are very rarely used in tropical countries.

- **The main mechanized tillage operations**

The main tillage operations discussed in this document are: ploughing, pseudo-tillage, subsoiling and seeding.

Ploughing:

It involves cutting and turning over a strip of land of varying width and depth.

Its essential role is to loosen the surface layer of the soil. By destroying the soil's surface crust and giving it a cloddy surface, we greatly improve water penetration, slow down evaporation, promote soil aeration and create a favorable environment for seed germination and plant development.

Ploughing is carried out by the following ploughs: share and mouldboard ploughs and disc ploughs. It is characterized by the depth of work. A distinction is made between :

---

13 R. Pirot (1998). La motorisation dans les cultures tropicales, found in Mémento de l'agronome.

- Light ploughing less than 10cm deep;

- Medium ploughing up to 15cm deep;

- Deep ploughing more than 20 to 25 cm deep.

The types of ploughing found in tropical zones are :

- Flat ploughing ;

- Plank ploughing and

- Ridge ploughing

Flat ploughing is the most common in tropical regions. It is carried out by animal traction or by motorization (tractor).

The advantages of ploughing in tropical regions :

- Soil loosening, following the turning of the soil, which causes it to break up into larger or smaller pieces that are then subjected to the action of the climate;

- The increase in the exchange surface of the soil with the atmosphere and the reaction of the lacunar network, favourable to the circulation of air and water in the soil and to the life of soil micro-organisms, in particular, there is an improvement in the infiltration and storage capacity of water;

- Destroying weeds by burial;

- Incorporation of organic matter (plant debris, manure, etc.);

- Erosion control.

The disadvantages of ploughing in tropical regions :

- Ploughing is carried out over a short period of time due to climatic conditions;

- If the soil is too dry, it is very difficult for the plough to penetrate and the pulling effort increases exaggeratedly;

- When the ground is too wet, however, the tractor's progress is severely hampered by a lack of grip (wheel spin).

In both cases, the improvement in structure is only imperfect and ploughing is slow.

- Ploughing is very energy-intensive: in fact, it's the cultivation operation that requires the greatest tractive effort.

Subsoiling:

It consists in working the soil at depth, without turning it over. Subsoiling is generally not carried out systematically every year (3 to 4 years at least).

It is generally carried out in the dry season, before ploughing, and as deeply as possible. The minimum depth is 40cm, and can be as much as 90cm.

The purpose of subsoiling is to :

- Breaking up any ploughing that may have taken place and shaking the ground deep down;

- Loosen the soil;

- Improve the circulation of water, air and fertilizing elements;

- Reduce the risk of rain erosion in rainy regions;

- Increase water reserves in regions with low rainfall. Subsoiling avoids the need for ripping ploughs, which run the risk of bringing sterile layers of soil to the surface and require enormous power and therefore very large traction equipment, which is not necessarily justified for subsequent work.

Subsoiling is carried out using an instrument called a subsoiler.

Subsoiling can be seen as a complementary operation to ploughing. It contributes to improving the profitability of agricultural production.

Subsoiling can be carried out using two different techniques:

- Follow the plough with an independent digging claw with 1 or 3 teeth, working the soil to a maximum depth of 40cm. This system, applicable to farms with limited means of traction, has the disadvantage of a double passage (compaction) and leads to the crushing of the excavated part.

- Subsoiling is carried out after stubble cultivation, when the soil is sufficiently moist. The subsoiler, which normally works at a depth of 0.5m or more, requires considerable traction. The furrows should be about 0.80cm apart. The operation is then completed by ordinary ploughing.

In Benin, tillage is not well known

Pseudo-labours

Their purpose is to work the soil at medium depth. It involves superficial loosening of the soil (5 to 15cm). Pseudo-ploughing can be carried out using a variety of implements, whose conditions of use vary greatly according to soil and weather conditions.

Some pseudo-ploughing tools.

In addition to loosening the soil, pseudo-tillage has other objectives:

a.   When it comes to preparing the soil for sowing, pseudo-tillage is an intermediate stage between ploughing and the superficial loosening work to be carried out at a later date. Pseudo-tilling consists mainly in bringing the clods left by ploughing up to the surface, in order to break them up, and in bringing the fine soil down to depth to avoid any discontinuity. But this practice has the disadvantage of leaving the soil excessively lifted, and requires subsequent compaction.

b.   Shallow burial of fertilizers

c.   Meadow clearing. This involves destroying the vegetation in the meadow, loosening the surface to allow rainwater to penetrate, and initiating the decomposition of plant debris. Tools such as scarifiers and disk sprayers are recommended for this type of work.

d.   Field maintenance using a scarifier or divider. This operation provides both superficial aeration of the soil and partial destruction of weak-rooted weeds.

e.   Stubble ploughing by scarifier, cultivator or disc sprayer.

■ Sowing

Sowing is one of the most important tasks, along with soil preparation. The timing and manner of sowing have a decisive influence on plant emergence, growth, maintenance and, ultimately, yield. Mechanically, seeding is carried out by an instrument called a "seeder". There are several types. There are several advantages to using a seeder:

- speed (time savings)

- regular plant spacing (good density)

- seed burial at an adjustable depth, - seed savings of around 30%.

-    **Mechanical harvesting**

Unlike other cultivation operations, for which tools can be used for most plant productions, mechanical harvesting requires specific equipment for each crop: the machine will be entirely different depending on whether you're pulling tubers, harvesting cereals or lifting peanuts. According to the agricultural machinery course Volume 1 (EDUCATEL), there are :

- For grain harvesting (mower, flail mower, binder harvester, merry-go-round thresher)

-   For tuber harvesting (potato harvester).

Mechanization is more than just a technical advance, it's also part of social evolution.

3-2 Mechanization: part of the social evolution of workers

When it comes to organizing work and work sites, animal traction helps solve certain bottlenecks by reducing peak workloads. It reduces the arduousness of work and raises the farmer's standard of living. It increases the motivation of young workers, helping to maintain village development dynamics. (G. Le Thiec, 1996)

The mechanized farmer is not only relieved of the burden of hard labor, but the increased productivity of land and labor also means higher yields and the possibility of extending the area under cultivation, with the same amount of labor.

He therefore sees his income increase and his family's standard of living improve: food is available, the children's education and the family's health are assured, housing is available, financial participation in the various family ceremonies is paid for regularly, and a means of transport (motorcycle, bicycle, etc.) is available.

It's true that all this comes with new constraints (housing, training, feeding and caring for the animals, tool maintenance, etc.), but it's rewarded by all the advantages mentioned above, to which we must add the production of manure and the transport of inputs and crops, for example.

A new class of wealthy men was created through the use of agricultural mechanization.

Whether at national or continental level, mechanization represents a decisive step in the modernization of agriculture.

IV- <u>SOIL FERTILIZATION AND PROTECTION OF PLANTS: CERTAIN FACTORS IN PRODUCTION</u>

Soil fertilization and plant protection are the keys to economic and sustainable agricultural production. In tropical regions, soil fertility declines from year to year as a result of poor farming practices such as lack of rotation and bush fires. Yields, meanwhile, are catastrophic due to damage caused by pests and parasites (such as locusts, caterpillars, striga, etc.) on crops. In order to solve these problems, soil fertilization and plant protection (PV) are necessary.

4-1 Soil fertilization

One of the main agronomic problems in tropical regions is maintaining soil fertility. The solution to this problem is not obvious, and is often closely linked to complex socio-economic issues. Nevertheless, informing farmers about soil fertilization methods is a necessary step.

Fertilization can be defined as all techniques that bring or maintain soil fertility at its highest level. Fertility is linked to the environment and the plant. It is the capacity of a soil to make a plant productive.

The notions of soil fertility and fertilization can be approached by analogy with those of human health and nutrition.

They must be analyzed in the context of other production factors, such as available technology and manpower. For a given region, the study of soil fertility and fertilization is therefore inseparable from the analysis of agronomic conditions, population dynamics and the effects of agricultural policy on agrarian systems (Piéri, 1982).

According to Boiffin and Sébillote (1982), the more general notion of the fertility of an environment can be likened to a judgment made about the functioning of a biological system whose interactive components (soil, climate and plant) are subject to technical, economic, social and historical determinants. Two types of fertility can be distinguished:

■ Natural fertility

It is related to the stage of degradation of soil minerals, which releases more or less exchangeable bases.

The overall level of fertility is closely linked to the presence of organic matter in the soil. Organic matter includes humus, which nourishes bacteria, and stable humus, which is a reservoir of nutrients for plants. This matter has a high buffering capacity and contributes to the formation of the clay-humus complex.

■ Acquired fertility

Acquired fertility is the result of man's work. This can be done in two ways:

1- By eliminating the negative factors that reduce fertility (in a soil with a perched water table in the rainy season, the first improvement to be made is chaining);

2- By bringing the soil to a high level of fertility, taking into account both the physical fertility (texture and structure) and the chemical fertility (mineralogical nature) of the soil.

The physical properties of the soil determine the effectiveness of fertilizers, and therefore fertility.

Taking into account the natural fertility of the soil in a given region and at a given time, soil fertilization aims to sustainably increase agricultural production. In tropical regions in general, and Africa in particular, soil fertility is low. As a result, crop yields are low. Apart from insufficient and/or poorly distributed rainfall, the causes of this drop in yields lie in inappropriate fertilization for soils that are naturally poor in organic and mineral matter, on the one hand, and unsuitable cultivation techniques, on the other.

Rabezandrina (1986) argues that, rather than defining agriculture as "the art of extracting maximum profit from the soil while maintaining its fertility", it should be conceived as "the art of conserving soil fertility, putting immediate profit in the background, which until now has prevailed in the fertilization approach, which most often consists of applying only mineral fertilizer to the soil in doses that result in negative mineral balances. So, in a forward-looking vision of agriculture, maintaining soil fertility must remain the primary concern of all those involved.

When it comes to improving soil fertility, especially in tropical regions, the addition of organic matter is essential. The effect of this organic matter on plant behavior is assessed not only in terms of the supply of nutrients, but also on the basis of its influence on the physical, chemical and biological properties of the soil, which in turn affect plant development.

Soil organic matter affects crop growth and development both directly, by making available to the plant the nutrients it releases, and indirectly, by improving the physical, chemical and biological components of soil fertility.

The action of organic matter on plants can be summarized as follows:

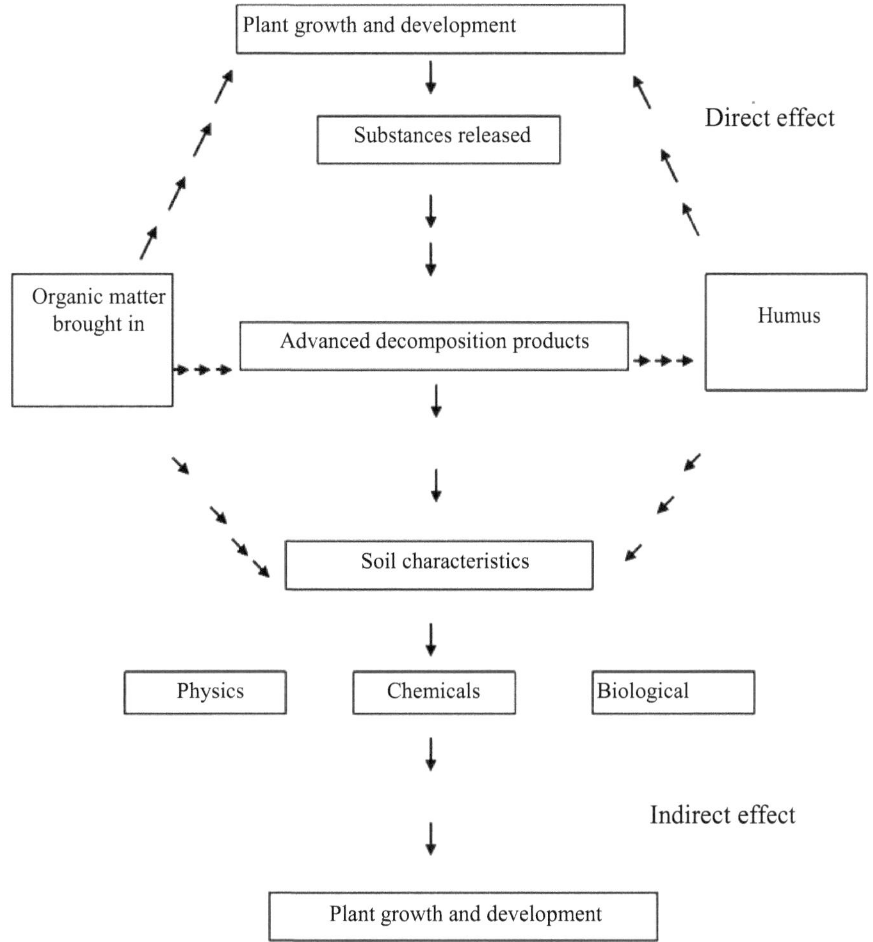

<u>Diagram</u> N°3: The decomposition of organic matter and its effects on crop growth and development.

Improved soil structure results from the presence of more and better organic matter. The same applies to improving the stability of the aggregates formed, water retention capacity, cation exchange capacity and reducing the soil's phosphorus-binding capacity. In addition, microbial activity in the soil also depends on the presence of organic matter. As an indicator of soil quality, soil organic matter is characterized quantitatively by the content of organic residues of plant and/or animal origin (dead leaves, grass clippings, household refuse, animal droppings, etc.) and qualitatively by the chemical and biochemical composition of these residues. Soil can be fertilized with organic matter in the following ways:

- Leave crop residues on the field, or park animals on the field, or install improver plants (e.g. *muccuna sp*) on the field, then plough the field, sealing off all the aforementioned elements with earth.

- Make the compost that will be spread in the field before ploughing, using the residues.

The practice of late bushfires, the use of crop residues for feeding domestic animals and abusive deforestation are factors that limit the production of organic matter on generally poor tropical soils.

To increase the productivity of these soils, it has been recommended to combine mineral fertilizers with organic matter. These are commercial artificial fertilizers, mainly nitrogen, phosphate and potassium fertilizers.

These fertilizers are delivered to plants in liquid form, but usually in solid (granulated) form.

When you want to fertilize a soil, i.e. apply fertilizers to optimize yields, you can't just do it at random: you have to respect certain principles.

4-1-1 Principles governing the use of fertilizers

The principles of fertilization are based on three (3) fundamental laws:

a- Restitution law

All the fertilizing elements removed by harvesting must be returned to the soil, otherwise it will be depleted.

This law recommends that manure be applied to the soil to match exports.

There are mechanisms that tend to make effective soil losses greater than crop exports:

- Leaching: bare soils and weeded crops are more sensitive to nutrient losses through chaining than grasslands.

Leaching is particularly significant for the following elements: nitrogen (N), potassium (K), magnesium (Mg), sodium (Na), calcium (Ca) and trace elements. However, losses are greater in light soils than in clay soils.

- Soil erosion: it removes the reservoir of fertilizing elements and the fertilizing elements.

- Retrogradation, blocking: P205, for example, precipitates in the form of calcium phosphates in very calcareous environments. Small quantities of potassium can also be retrograded (fixed to the crystalline lattice of clays).

- Luxury consumption: if N and K are provided too generously, there is a risk of exaggerated losses and exports.

In addition, there are natural mechanisms that compensate for losses and exports. These are essentially :

- Nitrogen supply through the work of microbes (nitrification, humus, atmospheric nitrogen fixation by legume rhizobia).

- Rainstorms bringing ammoniacal nitrogen

- Progressive release of insoluble mineral reserves through chemical and biological attack on soil minerals)

Let's remember that, rather than returning to the soil what's gone, it's better to make advances to the soil that will cover the needs of the future crop.

This means anticipating the needs of the future harvest. The first advance is the restitution of residues from the previous harvest.

b- Law of minimum or limiting factors

Yield depends above all on the nutrient that is in the lowest proportion to the plant's needs.

Most often, it's nitrogen that's in short supply. However, this is not the only limiting factor: it can also be other fertilizing elements, lack or excess of water, climate, variety, the presence of disease or pests, etc.

So, all growth factors can be limiting factors. Growth factors are generally interdependent: they interact with each other.

c- Law of less than proportional returns

When increasing doses of a fertilizing element are applied to a crop, the yield increase obtained becomes smaller and smaller as the quantities applied increase. In other words, an excess of one element in the soil reduces the effectiveness of the other elements, lowering crop yields.

Each factor can only produce its full effect in the presence of a sufficient quantity of all the others.

What's more, the maximum dose is not necessarily an economically profitable one: at some point, the extra fertilizer cost is no longer covered by the extra yield obtained.

4-2 Plant protection

The main aim of plant protection is to combat plant pests and the destructive action of man. Indeed, the importance of plant cover is a determining factor in rainfall and soil fertility, and therefore in agriculture. In this chapter, we will focus on the protection of crops and stored harvests, which are major problems currently facing agriculture in tropical countries.

Some crop pests are :

- Insects: Seven or eight important insect species can be found on the same plant;

- Parasitic fungi ;

- Viruses or bacteria;

- Birds;

- Rodents;

- Striga (parasitic plant).

All these enemies aim to make the plant sick, reduce or destroy its production, or kill it outright.

Among these numerous enemies, some are frequently found on crops and stored goods[14].

■ Culture level

Seedling sprouting: this is caused by various soil fungi, sometimes acting in combination. They can attack germinating seeds, resulting in poor seedling emergence. As a result, seedlings wilt, turn yellow and die, sometimes collapsing onto the ground.

The microorganisms responsible are present in most cultivated soils and are potentially harmful to many crops (beans, cotton, rice, etc.).

Fumaginia: this is a group of saprophytic fungi that cover the aerial parts of plants with a black, soot-like film. Scraping away this film reveals unaltered underlying plant tissue, a sign that the fungus is not a plant parasite.

When dense, fumagines interfere with the plant's chlorophyll and respiratory functions. Hot, humid weather and excessive shade encourage their development. They considerably reduce crop yields.

Cutworms (lepidoptera): they hide in the superficial layers of the soil. They only emerge at nightfall to gnaw on the collars of seedlings, which wither and dry up. A single individual can damage several plants. The presence of weeds, piles of garbage or plant debris in fields encourages cutworm attacks. Cutworms are polyphagous, attacking cotton, tobacco, coffee, beans, peanuts and potatoes, digging into their tubers.

White grubs (beetles): Both adults and larvae live in the superficial layers of the soil, preferably in moist soils rich in organic matter. Adults cause damage by gnawing on the crowns of seedlings, or by digging cavities in the roots of older plants (in the case

---

14 The range of pests and diseases cited is taken from "Maladies et ravageurs des cultures de la région des grands lacs d'Afrique centrale" (A. Autrique and D. Perreaux, 1989).

of maize) below ground level. Attacked plants spill onto the ground and dry out. These beetles attack maize and rice crops, but can also cause damage to beans, potatoes, etc. The larvae feed on plant debris on the roots, particularly of grasses.

Legionnaires' caterpillars (lepidopteran)*:* these eat away at the leaves of cultivated grasses (maize, sorghum, rice, wheat, etc.) and volunteer grasses, preferring to attack the early stages of plant development. Very large and dangerous larval populations can cause serious damage to young cereals. Attacks result from moth migrations. Wind and rainstorms play a major role in the migration of these populations from one region to another, sometimes far away.

Crickets (insects): there are generally two types:

- *The stink locust*: the lava and adults of this locust often live in large flocks and can cause considerable damage to crops. In particular, they gnaw on the leaves of cassava, cotton and young coffee plants.

- *nomadic locusts*: nomadic locusts can migrate from one country to another. Bands often numbering several thousand individuals can cause considerable damage.

In the region where they settle, they feed primarily on cultivated grasses (maize, sorghum, rice, wheat, etc.). Other cultivated plants such as cotton, bananas and palms are also attacked.

Crickets*:* Crickets cause damage during the night in nurseries and young plantations of coffee, tea, cotton, tobacco and yams. The larvae and adults use their strong mandibles to sever the stems of young plants and carry them off to their holes. Where crickets are numerous, they considerably reduce the number of plants lifted.

millipedes: live in the surface layer of the soil. They feed on decomposing plant or animal matter, but also gnaw on germinating seeds, seedling crowns, roots or potato tubers. Their proliferation can seriously reduce crop emergence (beans, peanuts, etc.). Millipede damage is enormous during the rainy season.

Root-knot nematodes: the larvae penetrate the roots and attach themselves close to the vascular zone, causing root-knot swellings in which they continue their development to become adults. The shape, size and appearance of galls vary according to their age, number, host plant, extent of attack and environmental conditions. In the event of heavy infestation, roots may be reduced.

The reduced root system and subsequent metabolic disorders result in poor plant development and a progressive drop in yield.

Weeds: these are the unwanted herbaceous vegetation on a farm. These weeds compete for food in the soil with established crops such as maize, cotton, beans, groundnuts, rice, etc. Generally, these weeds take over the crops. They slow crop growth and reduce productivity. There are several species. Take striga, for example. This species is the most formidable.

Weeds are one of the main biological constraints affecting food production worldwide. In tropical zones, the loss is estimated at 25% (Memento d'agronome page 669).

They also contribute to the high cost of labor time. In Togo, for example, weeding and ridging cotton crops account for 18% to 42% of the total time spent on the crop.

- **Stored food level**

Cereal moth (lepidoptera): a small, light-brown moth. It lays its eggs on grain. Infestation can begin in the field. As soon as it hatches, the caterpillar penetrates a grain and partially eviscerates it as it develops. Alucite can cause considerable damage to corn and sorghum stored as bare cobs or panicles, respectively. It also attacks wheat and rice grains.

Weevils (beetles): they reproduce mainly at the expense of cereal grains: corn, sorghum, rice, wheat, etc. The damage they cause and their control methods are the same as for alucite.

Corn borer (beetle): adults bore holes into corn cobs, even through the spathes. Eggs are deposited in cavities in the kernels. After hatching, the larvae feed and develop inside the kernels.

The capuchin beetle can cause considerable damage to maize, especially when stored on the cob. Infestations can begin in fields before harvest. The preventive measures recommended against the cereal leaf beetle also apply to the greater capuchin beetle.

Bean bruchids (insects): the female lays her eggs in ripe pods in the field, or deposits them on stored beans. After hatching, the larva penetrates the seed and continues its development by burrowing into it. The larvae of a single female can destroy an entire stock.

There are many different sources of plant infestation, but one of the most important is the way the parasites follow their course. They survive on wild hosts, without necessarily producing symptoms in the soil, on plant reproduction or multiplication. Some insects and fungi are transported over long distances by the wind, and the sources of infestation are also sometimes a long way from the threatened crop.

Locally, parasites are often spread by rain, run-off or irrigation water, or by simple contact between plants. Animal parasites are generally self-propelled, and are also vectors of various pathogens. Humans are a significant factor in the spread of pathogens, through the transport of infested or contaminated plant material, certain cultivation practices or simply by walking through a crop.

The fight against plant diseases and pests is first and foremost preventive and cultural.

To prevent the introduction of crop pests (insects, nematodes, bacteria, pathogenic fungi, viruses, acarids) into a country or group of countries, phytosanitary legislation has defined a number of principles.

4-2-1 Principles governing the use of plant protection products.

These principles include

a- **Prohibition,** i.e. a complete ban on imports ;

b- **Import permits**: the prior import permit system generally applies to a limited number of plants, for which a request for authorization must be submitted to the local plant protection service prior to import;

c- **The health certificate: this is** a document issued by the department responsible in the country of origin, stating that the material is in a satisfactory state of health. In some cases, it may be replaced by a simple certificate of origin;

d- **Arrival inspection**: to check declarations, declared certificates and the condition of the product itself;

e- **Disinfection**: this operation is sometimes carried out before the goods are loaded. It can also be carried out after inspection on landing;

f- **Quarantine**: when the above precautions are deemed insufficient to avoid the risk of introducing a new disease, we can resort to quarantine cultivation, which must be carried out in an approved station offering sufficient guarantees.

In tropical regions, crops and stored goods are subject to attacks of all kinds, due to the sheer number of enemies. If cultivated plants are to be in good condition to produce quality fruit that is highly competitive on the commercial market, these pests need to be controlled. In fact, the ideal, well-thought-out approach is one that incorporates all the means currently available to combat these pests: cultivation techniques, resistant varieties, biological control and rational use of plant protection products.

However, the implementation of crop protection in the strict sense of the term still presents enormous difficulties, mainly due to the lack of financial resources needed to purchase products and the lack of technical training for farmers. It was these difficulties that gave rise to the idea of giving priority, in the relatively pristine ecosystems of the intertropics, to the maintenance of biological balances and the development of biological control[15] .

4-3 Description of some results obtained from the use of fertilizers and phytosanitary products

According to M. ROTH, 1983, comparing the yields of "well-treated" plantations with those of poorly maintained plantations in terms of both pesticide treatments and

---

15 M. ROTH, 1983. Crop protection in tropical regions. B2-GEOG.-40. B.T.I 379/381 - 1983

manuring, the figures are suggestive.

In traditional cultivation, maize produces 500 to 800 kg/ha for 3 to 4 tonnes in field cultivation; rice yields can vary from 8 to 40 q/ha (the only insecticide treatment improves this yield by 10 q/ha); sorghum can produce 12 q/ha in improved cultivation compared with 6 in traditional cultivation; cowpeas often produce less than 200 kg/ha in traditional cultivation and 1000 to 1500 kg/ha in pure cultivation protected against parasites (with fertilization, it is true).

As far as stored foodstuffs are concerned, precise figures can of course be provided.

In Senegal, losses for groundnuts were recently estimated at 30 - 40%; in Cameroon, in traditional granaries, 46% of stored maize is destroyed after 6 months in the case of local composites, and almost 100% for newly introduced varieties. In fact, of the one and a half billion tonnes of perishable foodstuffs produced worldwide, 40% of the quantities not kept cold (by drying, salting, cooking or smoking) are destroyed.

V- **THE BASICS OF RUNNING A PROFITABLE FARM IN A TROPICAL REGION**

The profitability of a farm, like any other business, depends on how it is managed. In tropical regions, many farmers fail to achieve their campaign objectives because of poor farm management.

The role of management, then, is to ensure that a farm's activities are a success. As such, it provides solutions to the problems facing the operation.

The search for solutions to the difficulties encountered by farmers can be subdivided into four phases (CRET, Techniques Américaines No. 82):

- Define the problem.

- Design various measures to solve the problem.

- Select the solution that best meets your business objectives.

- Evaluate the effectiveness of the chosen process.

In addition, to ensure the success of his business, the farmer performs the following main functions:

a- Forecasting :

For a farm to be well managed and profitable, it all starts with forecasting. Forecasting is an activity of reflection, judgment and decision-making. Forecasting means preparing for tasks, not carrying them out. It enables the realization of events which, without its intervention, would not occur.

A good forecast is based on an analysis of the farm's strengths and weaknesses, in order to set objectives for the season. At this stage, the farmer reflects and makes decisions by answering the following questions:

- What are the different crops to grow this year?

- On which plots to grow crops?

- What human, material and financial resources are available?

- How much area to plant per crop?

- Is manpower available?

- What is the expected yield per crop?

- What is the estimated cost of the various operations?

- How profitable is the operation?

- Etc.

Answering all these questions leads to the development of a campaign plan and a projected operating account.

Forecasting is of paramount importance in farm management, because what's the point of planting land if you can't maintain it?

Once you've decided what you want to do, and what you're going to do it with, it's essential to ask yourself "how to do it", hence the need for work organization.

b- Work organization :

It's all very well to bring together all the resources needed to manage a farm, but if the work organization is poor, the farm will never be profitable.

Planning allows us to organize the means available and those to be found to carry out the actions identified and considered viable and feasible, in order to achieve the objectives (making the operation profitable) previously set within a given timeframe.

Good organization of work requires that the people involved are qualified to do their job. This calls for training and exchanges of experience on farm management techniques.

All in all, organization calls for good planning that addresses the following questions:

- Who does what on the farm?

- Why? Why?

- Where? When? With how many people?

c- Management

Directing is a management function concerned with the day-to-day, detailed execution of the various tasks to be carried out in the operation. Directing means giving orders to the operating staff, telling them what to do, when to do it, and how to do it. Above all, leadership requires the ability to draw lines of conduct and keep an overview.

It is the responsibility of the person in charge of the operation to explain the tasks assigned to staff, using all appropriate teaching methods.

d- Coordination

Coordination means integrating activities, personnel, facilities and equipment into a single whole, in order to achieve a seamless, satisfactory operating cycle. Coordination therefore means ensuring that each operational unit assists the others to the best of its ability, without risking conflict.

e- Control

Control consists firstly in checking the results actually achieved against the previously established plan, to ensure that they are carried out according to plan, and secondly in adopting corrective measures to avoid undesirable results.

The control process comprises three stages:

- Results assessment by the operations manager

- Measuring results

- Modification of plans that are not materialized as planned.

VI- <u>THE NEED FOR HYGIENE AND LABOR WELFARE IN AGRICULTURAL PRODUCTION IN THE TROPICS.</u>

Hygiene is the set of rules we follow and the precautions we take to protect our own health or that of a community.

For people, well-being means easy access to sufficient, quality food, decent housing and clothing, health services, education, water and a healthy living environment. Hygiene therefore plays an important role in creating the conditions for human well-being.

In tropical regions, agricultural production relies essentially on human energy. In these conditions, the search for a quality workforce is essential to promote agriculture. However, unhealthy environments and outdated practices characterize certain populations in countries of the tropical region. As a result, the health of the workforce in these countries is a key issue. Farming operations are poorly mechanized, and the precarious health of the workforce has a negative impact on farm productivity.

It is worth noting that in some villages :

* ***Non-observance of good hygiene***

Piles of garbage and animal droppings fill the courtyards of houses with people. People drink contaminated water and eat meals of dubious quality, as well as the meat of sick or dead animals. They wash with poor-quality water, and clothing is irregularly laundered. Suspect sexual relations are frequent (source of AIDS, SYPHILIS...) and teeth are poorly maintained (source of tooth decay). In fact, hygiene in general is poorly observed. This situation can be explained by the lack of Information, Education and Communication (IEC) on the one hand, and the fatalism of the population on the other.

*   ***The well-being of the population not assured***

Some populations don't have enough to eat, and don't get good-quality meals either. They have difficult access to education and health services. Their homes are exposed to the vagaries of the weather (rain, strong winds, etc.). Their well-being is not well assured. The causes of this situation are poverty and insufficient Information, Education and Communication (IEC).

Given all the above, the health of the people who make up the workforce in regional agricultural production is precarious.

Good health is an important element in the promotion of any activity. Consequently, the hygiene and well-being of the workforce in agricultural production in tropical regions is a necessity. This means that :

-    IEC sessions on food, body, clothing, dental, sexual and environmental hygiene are organized for the population;

-    populations have easy access to healthcare;

-    populations adopt good hygiene practices.

VII- <u>SPECIFIC CASE OF MAIZE AND COTTON CULTIVATION</u>

In this part of the summary, we will look at aspects relating to labor, improvement factors and the marketing of maize and cotton.

Reasons for choosing these crops

■   **Corn**

Among the food crops produced in Benin, maize was chosen because it is not only highly developed in tropical regions, but is also the subject of technical support to producers in northern Benin by a number of players.

In terms of food, maize is highly appreciated by people living in the tropics in general, and in the Republic of Benin in particular, as it can be eaten in many forms: grilled, paste, cake, couscous, ground porridge, bran and grain for animals...

In economic terms, maize is currently sold like any other cash crop. Whether sold raw or processed, it is a source of monetary income for the local population.

In terms of cultivation, it's true that maize is nutrient-intensive and quickly exhausts the soil, but it's not difficult to grow if the technical itineraries are well known and applied.

On the conservation front, there are many endogenous solutions and very suitable plant protection products that can enable grain corn to retain its intrinsic quality for months without any parasitic attack.

- **Cotton**

As far as industrial cultivation is concerned, cotton, the main cash crop, was chosen because it is grown almost everywhere in warm countries and is "Benin's white gold".

Commercially speaking, the cotton market has always been available. It's true that the price per kilogram varies with each season, but cotton has never remained unsold. It is Benin's leading export product. Among export products, cotton brings in the most money for Beninese producers. This is why the President of the Republic of Benin has made the promotion of this crop his top priority.

From an organizational point of view, the cotton industry is well organized in Benin, as it is everywhere else.

In terms of cultivation, the climate is favorable, inputs are available and, in some regions, the soil is also available and favorable.

Another reason for this choice is the wealth of experience we have gained in providing technical assistance to producers.

In order to successfully carry out the various cultivation operations for maize and cotton, good planning is essential.

7-1 Labor planning for maize and cotton cultivation :

The aim here is to distribute the various cultivation operations to be carried out by the workforce (from sowing to harvest processing) in time and space for the cultivation of maize and cotton. Soil preparation work is not taken into account in this planning.

For corn :

The corresponding labor schedule (Table N°1) concerns maize (variety TZB-SR) with a long cycle (4 months) and takes into account the reality of northern Benin. Operations are divided according to periods.

For cotton :

The corresponding timetable (Table No. 2) is part of an on-time sowing program in northern Benin.

Generally speaking, family labor is more in demand for field work than hired labor.

Table 1: Labour planning for maize cultivation

N°	Cultivation operations	J	F	M	A	M	J	J	O	S	O	N	D	Type of labor	Quantity of labour per hectare
	Corn														
1.	Sowing						—							Family	4 men/day
2.	Herbicide treatment						—							Family	1 man/day
3.	1ᵉʳ NPK fertilizer spreading							—						Family	3 men/day
4.	Weeding						—	—						Salary	10 men/day
5.	2ᵉ spreading NPK fertilizer							—						Family	3 men/day
6.	Spreading urea fertilizer								—					Family	2 men/day
7.	Sarclo butting								—					Salary	10 men/day
8.	Weed control									—				Family	5 men/day
9.	Harvesting and transport										—			Family	15 men/day
10.	Drying (for 03 days)										—			Family	3 men/day
11.	Attic storage										—			Family	3 men/day
12.	Ginning (mechanized)											—		Salary	5 men/day
13.	Phytosanitary treatment of crops											—		Family	2 men/day

N°	Cultivation operations	J	F	M	A	M	J	J	O	S	O	N	D	Type of labor	Quantity of labour per hectare
14.	Storage (bagging, warehousing)												—	Family	3 men/day

Table n°2: Labor planning for cotton cultivation

N°	Cultivation operations	J	F	M	A	M	J	J	O	S	O	N	D	Type of labor	Quantity of labour per hectare
	Cotton														
15.	Sowing						—							Family	4 men/day
16.	Herbicide						—							Family	1 man/day
17.	Wedding							—						Family	3 men/day
18.	NPK fertilizer spreading							—						Family	4 men/day
19.	Weeding							—						Salary	10 men/day
20.	Spreading urea fertilizer							—						Family	2 men/day
21.	Sarclo-buttage							—						Salary	10 men/day
22.	1er phyto treatment								—					Family	1 man/day
23.	2e phyto treatment									—				Family	1 man/day
24.	Weed control									—				Family	5 men/day
25.	3e phyto treatment										—			Family	1 man/day
26.	4e phyto treatment										—			Family	1 man/day
27.	5e Phyto treatment											—		Family	1 man/day
28.	6e phyto treatment											—		Family	1 man/day
29.	Harvest												—	Salary	30 men/day

7-2 Factors likely to improve maize and cotton crops

These factors include :

a)   **Soil**: the soil on which corn and cotton are grown must be well-drained and fertile. These crops should be planted taking into account their place in the rotation. The soil must be ploughed flat and in good time, burying plant debris properly.

b)   **Water**: water is a vital element for animals and plants. For both crops, rainfall must not only be sufficient, but also well distributed. In northern Benin, for example, it takes an average of 1,200 mm of rainfall per year for rains to fall at intervals of around five days.

In regions with large irrigation dams (e.g. Burkina Faso), water must be supplied to the plants as and when needed.

c)   **Inputs**: inputs must be specific and of good quality.

- For maize, the variety must be improved, selected and recommended by an agricultural research department. Apart from organic fertilizer, which is applied on site, specific chemical fertilizers (e.g. PA) must be used. The herbicide used to control weeds must be of good quality and specific to maize: (Primagram, for example).

With regard to cotton, the variety to be sown must be chosen taking into account the soil types and climate of the area. The chemical fertilizer to be used should be that recommended by the relevant authorities (NPK and urea, for example).

To combat weeds, use cotodon, for example.

For the six (6) compulsory phytosanitary treatments (case of Benin) products such as: andosulfan, percal, decis ... must be used.

NB: The doses of inputs recommended by the competent authorities for a given plot of land must be respected: no more, no less.

d) **Capital (money)**: these days, farming without money is very difficult, if not impossible. Money is needed to buy inputs, to pay for plowing or motorized cultivation, to hire labor, and so on. Money helps to ensure that the various cultivation operations are carried out on time.

e) **Labor:** to improve these two crops, labor must be available. In tropical regions, rainfall is generally short-lived. Labour will enable the various cultivation operations to be carried out quickly and in good time. It will also enable fields to be properly maintained to improve yields.

7-3 Economic opportunities for improved marketing: Marketing agricultural products is the driving force behind agricultural development in tropical regions. For the farmer, it is a very decisive phase, as it determines the economic profitability of the campaign. Farmers' impressions of last year's marketing enable them to make decisions about the coming crop year.

Example: if marketing has been better and brought in a lot of money, farmers opt for good field maintenance.

The economic profitability of marketing depends on several factors:

a) **Product price formation :**

Corn

Maize is both a subsistence food crop and a cash crop. In most countries in the tropical region, there is virtually no policy for price formation for this product by the players involved. Each trader goes down to the villages and buys the product according to his financial capacity and negotiating technique. In Benin, for example, some traders in the same village buy 25 kg of maize for 1,500 CFA francs, while others buy it for 1,250 CFA francs in December and January.

Cotton

Cotton marketing is well organized in tropical countries. In Benin, for example, the price per kilo is set by the government in collaboration with producers. But producers are always teased.

In general, farmers are less involved in price formation. For marketing to be economically profitable for all players, it would be necessary for :

- all stakeholders are involved in product pricing;

- the application of the selected prices is imperatively respected by the players concerned;

- sanctions are imposed on those who have broken the law.

b) **The large number of intermediaries :**

The marketing chain is long in some tropical regions because of the number of intermediaries. A single product can have up to five (5) intermediaries. In Benin, for example, a single trader has several collectors at different stages of the maize marketing chain. Urban collectors entrust money to village collectors, who in turn entrust money to farm collectors. The longer the chain, the less the producers earn. In order to reduce this chain and improve economic opportunities through better marketing, we need to :

- agricultural sectors are well organized;

- product bundling is promoted.

c) **Transport difficulties :**

In some tropical regions, growers are far from marketing areas. Roads are faulty. Merchants lose a lot of time and money before reaching producers. After all the economic calculations, traders are obliged to reduce the cost of their products. It would be necessary for :

- governments rehabilitate rural tracks

- marketing zones are closer to producers.

In view of the above, it has to be said that marketing channels are not very rational and are controlled by a few, with no possible alternatives. The producer is therefore dependent on them.

The large number of intermediaries penalizes the producer, as his price will be reduced accordingly. The lack of secure outlets, due to the lack of permanent road infrastructures, makes the marketing of agricultural produce in production zones uncertain.

Part Two: FIELD RESULTS AND DISCUSSIONS

This section presents the results of the field survey, discusses them with those of the literature review and draws conclusions.

The field survey was carried out in the Borgou/Alibori department (northern Benin), specifically in the communes of Tchaourou, Parakou, Nikki and Banikoara. The resource persons involved in the survey were farmers, agricultural extension workers and engineers.

From the summary of questions on the survey form, the following main points emerge:

- soils and soil improvement ;

- The main crops and their importance ;

- Ground works;

- Support provided to farmers by extension workers and engineers. The results will be presented according to these different points.

1- *soils and soil improvement*

Farmers classify soils according to color. The main soils observed are :

- *black soils*

They are characterized by fine, humus-rich soil. Rotting vegetation is found on these soils. The main crops grown on these soils are sorghum, maize, yam, cotton, shea, néré and cashew. These are certainly tropical ferruginous soils found in the Sudanian zone of northern Benin.

- *gray floors ;*

The presence of clay on these soils is noted. Crops such as maize, sorghum, millet and cotton are grown in this zone. These soils are found at Banikoara in the north-western Sudanian zone.

- *light sandy soils.*

On these soils, sand occupies a greater proportion. These soils are found throughout the North. Crops grown include groundnuts, cowpeas and cassava ....

Farmers select land for cultivation during the rainy season, specifically in August-September. Good land is characterized by the height, greenness and species of grasses, and by the density of trees. For head crops, for example, good soils are those whose herbaceous vegetation is dominated by <u>Penissetum purpereum</u>. Poor soils, on the other hand, are characterized by the presence of dogtooth <u>(Imperata cylindrica)</u> and striga (<u>Striga hermonthica)</u>.

The farming method used by farmers is to remove almost all the trees, leaving the soil bare. For this practice, farmers state that "shading trees does not allow crops to develop properly and produce". Indeed, shading large trees deprives low-growing crops of light, and photosynthesis is poorly ensured. But the risk is that this method exposes soils to erosion in all its forms, and their stability is not guaranteed. Faced with this no less important risk, precautions need to be taken to ensure the stability of these soils. In addition to the stability proposals set out in the documentation, we need to :

* avoid :

- 	Cropping systems that deplete organic matter ;

- 	To leave the ground bare;

- 	Late bush fires destroy microorganisms and other soil animals.

* promote :

- 	Organic matter inputs, for example by leaving crop residues on site or cultivating green manure plants such as mucuna;

- 	Cultivation of plants with good soil cover ;

- 	Growing trees in fields such as Gliricidia sepium.

In addition, to improve the agricultural value of soils, farmers practice :

fallow land

The fallow period is the determining factor in the natural vegetation of soil fertility.

The value of the system is based on the ratio between crop duration and fallow period. The fallow practiced by farmers lasts between 2 and 3 years, after which the soil is returned to cultivation. Farmers are aware of the short duration of this fallow, which contributes less to soil fertility. The reason they give is the shortage of arable land due to population growth.

Some are forced to shorten, or even abolish, the fallow period. As a result, too few plant nutrients can accumulate again or be retained, and soil fertility gradually declines.

The practice of fallowing, essential to ensure a minimum level of subsistence, still plays an important role for people living in the tropics.

In conclusion, a fallow should last between 5 and 10 years. In this way, it promotes good soil cover and protects the soil against all forms of erosion (rain, wind). After a few years, it gives the soil its natural fertility, thus reducing the use of increasingly expensive chemical fertilizers: a 50kg bag of NPK or urea fertilizer cost 11750FCFA in Benin in 2006.

Fallowing by WOURAHI Jean: farmer at Tchaourou (photo: MOUSSA Issiakou).

Crop rotation.

Growing the same crop on the same soil for several years contributes to a decline in soil fertility and encourages the development of certain harmful parasites. The practice of successive cropping helps to avoid such damage. Rotation is possible, whatever the size of the farm. Certain plots of land on the farm can be taken out of rotation, either permanently (orchard) or temporarily.

In the village of Sérékali, in the commune of Nikki, producers practice rotation according to the following table:

Year	Crops	Reason
1	Yam	Requires fertile soil and well loosened soil for future cultivation
2	Corn	Requires well-tilled, fertile soil
3	Cotton	Availability of mineral fertilizers to compensate for elements

		extracted by corn
4	Corn	Residual effects of cotton still available in the soil
5	Fallow land	Enables natural fertility

Table 3: Rotation used by SEREKALI farmers in NIKKI commune.

This is the practice taught by agricultural extension workers and engineers.

It should be noted that proper crop rotation is necessary for :

- Ensure a good chemical and humic balance in the soil. In this context, plants with taproots (e.g. cotton), which draw nutrients from deep down, should alternate with those with superficial roots (e.g. maize), which feed on the surface,

- Ensuring that certain crops are planted in good conditions.

- Controlling crop pests: this involves alternating plants from different families to limit the development of specific pests.

Setting up perennial and annual crops

The farmers we met do not plant perennial and annual crops with the aim of improving the agricultural value of the soil, but for their fruits.

In terms of perennial crops, fruit trees such as cashew and mango have been installed because of their economic purpose.

Compared with annual crops, after harvesting, the residues (tops, stalks, etc.) are entirely collected and used for animal feed, leaving the soil bare. In effect, the law of restitution is not respected. As a result, the soil is impoverished year after year and becomes useless for agriculture. A particular case of residue management was observed in Banikoara at the KESSE Abraham farm. This involves two ploughings prior to sowing, using a tractor-drawn cultivator. The first is done during the very first rains (end of April, beginning of May) and allows crop residues to be buried. Until then, nothing is sown. The second is carried out as soon as the rains begin (June). During the second ploughing, it is observed that the plant debris has rotted away. After this ploughing, the seed is sown and the farmer uses less mineral fertilizer.

To improve the agricultural value of soils through perennial and annual crops it would be desirable that :

-       the installation of plant improvers such as Muccuna utilis, Gliricidia sepium and Cajanus cajan.

-       the experience of Mr. KESSE Abraham (producer in Bouhanrou, commune of Banikoara) be implemented.

No overgrazing or bush fires.

The frequency of bush fires was noted during the survey. These fires not only deprive animals of their food and shelter, but also burn organic matter and kill the micro-organisms responsible for decomposing it. The soil becomes bare and dry. This exposes it to erosion. It also becomes poor and unprofitable for agriculture.

To correct this practice, it would be advisable :

-       back up awareness-raising with repressive measures;

-       installation of village bushfire monitoring committees ;

-       develop and implement participatory annual bushfire management plans.

2- the main crops and their importance

Farmers practise subsistence farming, which is poorly mechanized and dependent on the vagaries of the weather.

The main crops grown by farmers vary according to the area occupied, the volume of production and the monetary income generated. Following an exploratory survey carried out in nine (9) villages in the communes, the results are as follows: see table n°4.

Crops	Ranking of main crops by											
	Area occupied				Production				Income			
	Banikoara	Tchaourou	Parakou	Nikki	Banikoara	Tchaourou	Parakou	Nikki	Banikoara	Tchaourou	Parakou	Nikki

Corn	2	1	1	1	2	1	1	1	2	2	2	2
Cotton	1	3	3	2	1	3	3	2	1	3	3	1
Yam	5	2	4	3	4	4	5	3	3	4	5	3
Cashew tree	6	4	5	4	6	2	4	4	4	1	1	4
Cassava	4	5	3	5	5	5	2	5	5	5	4	5
Sorghum	3	6	6	6	3	6	6	6	6	6	6	6

Table 4: Ranking of main crops by area, production and cash income

Generally speaking, according to the survey results, maize comes first for food crops and cotton for cash crops. The position of these two crops in the ranking is one of the reasons why they were chosen to address the question of labor planning (see point 7).

In terms of crops as a whole, the following was observed:

* *Compared to selected seeds* :

There are agricultural research stations involved in seed multiplication and processing. These include

- INA Agricultural Research Center

- ALAFIAROU seed farm

Despite the existence of these research centers, a number of problems remain:

- Users do not take ownership of the results. As proof, research has discovered a method of multiplying yam seedlings which, according to producers, is a lengthy process that is immediately less profitable and requires more time and money for care.

- Although research is making commendable efforts to improve cereal crops, there is still a problem with disseminating the results. Research in Benin has identified the TSB-SR and DMR maize varieties as being adapted to specific areas of northern Benin. To date, although these seeds are highly productive, they are difficult for producers to access due to a lack of information and collaboration between research stations and farmer support structures. In addition, the high cost of improved seeds limits producers' access to them.

\-   It should be noted that the involvement of users of research results in the selection of improved seeds is low. At the same time, all voices agree on the fact that if a selection is not participatory, its results are not convincing.

Convincingly, all the growers interviewed recognize that the use of selected seed contributes to improving the quality and quantity of agricultural produce. For this seed to be truly used by farmers, it would be necessary to :

\-   Take farmers' real needs into account when implementing research projects

\-   Promote a participatory approach to the implementation of research projects, involving all stakeholders from project design to evaluation.

\-   Adopt a pricing policy for improved seeds that facilitates access for farmers

\-   Adopt a distribution policy that brings selected seeds closer to farmers

** In relation to fertilization :*

Only NPK and urea mineral fertilizers are used by farmers to fertilize soils, which are generally poor. Organic fertilizers such as manure, compost, crop residues, etc. are not applied to the fields. Farmers consider that the quantity of manure or compost to be applied to a hectare is very large and requires a lot of transport (plough or cart). They therefore prefer to buy mineral fertilizer, despite its relatively high cost (12,000 f cfa/50kg) and its scarcity on the market. Of all the crops grown, only cotton has specific fertilizers. The dose recommended by extension workers is 300kg of NPK and 50kg of urea for one hectare of cotton and maize. Farmers apply an average of 400 kg NPK and 100 kg urea per hectare. The average yield obtained is 1.200 T/ha for cotton and 3.5/ha for maize. A comparison of the two practices shows that farmers are over-fertilizing. This excess can pollute the soil and reduce biological activity. As a result, over the years, the soil's colloidal state declines and it is exposed to erosion in all its forms.

Improved soil structure is the result of the presence and quality of organic matter. So we need to :

-   combine organic matter and mineral fertilizer in a field to reduce fertilizer costs, enhance soil stability and significantly increase yields;

-   facilitate access to specific fertilizers for growers ;

-   promote compliance with advice from extension workers on the use of mineral fertilizers.

* In *relation to insects and diseases*:

Crops and stored goods are seriously attacked every year. One out of every two growers surveyed is a victim of attacks on both crops and stored goods every year.

In terms of crops, the main insects are caterpillars, locusts, bugs, termites, centipedes and crickets. The main disease is sudden death of plants, which can be caused by fungi.

For stored foodstuffs, it has been weevils, capuchins and molds that are caused by humidity and fungi.

To control crop pests, farmers use both traditional and modern (chemical) treatments.

Traditionally, they use neem filtrate to combat caterpillars and bedbugs, but only on small areas.

In the village of Gokanna (Tchaourou), for example, the farmers we met bury around ten fruits of a tree called "BOUEROU" in the local Baatonou language, over an area of one hectare, to keep termites away from the fields. This treatment does not kill the termites, but sends them away. For chemical treatment, the products regularly used are those (decis, andosulfan) designed to combat the enemies of the cotton plant. Growers are not aware of the phytosanitary products used on other crops.

To preserve foodstuffs (corn, sorghum, beans, etc.), farmers use ash, tobacco leaf and chili pepper. With this treatment, products can be preserved for up to six (6) months without being attacked. They also use chemical products such as sofagrain, actellic and phostoxin.

Farmers' constraints are :

- lack of knowledge about specific plant protection products ;

- the high cost of existing phytosanitary products;

- failure to control the use of existing phytosanitary products

Phytosanitary products, or pesticides, which are used to control pests of crops and stored foodstuffs, present dangers to the user, the consumer and the environment if they are not used judiciously. So, to minimize the risks, we need to :

- agricultural research is focusing more on biological control techniques;

- endogenous control techniques that have proved their worth are promoted and disseminated among the stakeholders concerned;

- Farmers are informed and trained on the advantages and disadvantages of phytosanitary products.

* *In relation to farm management* :

The head of the household is primarily responsible for managing the farm. He is the farm manager. He is the one who brings family members together to discuss the crops to be cultivated, the areas to be sown, the execution periods, ... All decisions taken are not transcribed onto paper, but remain verbal. The farm manager ensures that decisions are implemented. All cultivation operations (sowing, weeding, phytosanitary treatment, harvesting, etc.) are generally carried out by family labor. As part of farm management, farm managers are trained by projects and NGOs (Projet d'Appui à la Diversification des

Systems, for example) on drawing up and implementing a farm account. However, very few farmers apply what they have learned. It would be desirable to organize follow-up/advice sessions to reinforce the capacity of those trained, as the profitability of a farm depends on the quality of its management.

* *In relation to marketing* :

Apart from cotton, the marketing of other agricultural products is unorganized.

Producers find it difficult to sell their products. They are confronted with the problems of :

- lack of job opportunities ;

- low purchase price;

As a result of these problems, farmers' incomes are low and production is falling year on year. In fact, marketing is the driving force behind agricultural development in tropical regions. So, to promote this marketing, it would be desirable for :

- agricultural commodity chains are organized;

- Producers organize group sales in the villages;

3- soil preparation

Farming is traditional, with very little mechanization. This situation is due to a lack of money, which prevents farmers from having access to mechanization equipment. The tools regularly used are the hoe, the daba, the axe and the cutter. Harness farming and motorization are not widespread. In the commune of NIKKI, which has a population of 66,164[16] , there are 1,528 head of draught oxen, 3,395 complete ploughs, 300 carts and 02 working tractors. Most of the work in the fields (ploughing, sowing, weeding, ridging, harvesting, transporting produce, etc.) is carried out manually. As a result, farmers not only waste more time on small areas, but also fail to carry out cultivation operations on time.

Farmers declare: "the use of tractor-assisted cultivation allows us not only to diversify our agricultural activities, but also to obtain a better yield". This statement was confirmed after collecting results from 80 cotton and maize growers (see table below) in the commune of Nikki.

Working tools	Culture	Yield Medium	Observation

---

16 National census 2002

Rudimentary (hoe, daba, cutter, .)	Corn	1.5 T/ha	Hard work, takes time and a lot of people.
	Cotton	0.800 T/ha	
Mechanization (traction animal)	Corn	3.5 T/ha	Time-saving , lesspainfulbut expensive.
	Cotton	1,200 T/ha	

Table 5: Results of mechanization and rudimentary tools.

Corn field (top) and cotton field (bottom) of Mr. MOUROU Pascal: producer in Banikoara (photo: MOUSSA Issiakou)

The results of this table show the advantages and disadvantages of mechanization:

❖ *Advantages of mechanization*

1-   Increase in agricultural labour productivity :

Labor productivity is measured by the ratio between the quantity produced and the quantity of labor provided (number of men, for example). Thanks to mechanization, man is relieved of some of his hardest work: for the same output, he can work less, or conversely, if he has enough land, he can produce more with the same amount of work.

2-   Increased land productivity :

The farmer can concentrate certain tasks at the most favourable time (ploughing, sowing, harvesting); the work is done more efficiently (especially soil preparation, sowing, etc.), and certain tasks can be carried out by mechanization that previously seemed impossible (deep tillage, difficult-to-work land). He can also intensify cultivation by introducing advanced crops, such as green beans.

Finally, all of this helps to increase yields and improve product quality and therefore value.

3-   Changing habits and diversifying work :

Deeply traditionalist, farmers are often reluctant to change their way of life. Mechanization will upset his farming habits (grubbing-up, flat ploughing, sowing in rows, etc.) and lead him to take on new activities: animal care, tool maintenance, etc.

❖ *Disadvantages of mechanization*

The main difficulty holding back the expansion of farm mechanization is the lack of investment possibilities for farmers, given the high cost of mechanized tools and services. They need money to buy animals, equipment and even a tractor, which will undoubtedly cost a lot of money.

The farmer's meagre savings will often be insufficient for such expenses, and he will have to resort to credit.

On top of this, the farmer needs to take time out to attend training courses on animal training, for example, or on how to manage working tools... Driving the animals also becomes a problem, as children need to be sent to school.

From all the above, it has to be said that mechanization has relevant economic interests on the development of a nation, but when choices are not rationed, it can lead to social and environmental consequences.

The rational choice of techniques for agricultural mechanization must necessarily take into account the constraints of the environment, the plant and the socio-economic context.

- In relation to environmental and plant constraints:

The need to respect the crop calendar calls for the use of rapid means of installation, i.e. the use of animal traction or agricultural machinery.

What's more, improper use of motorization (deep ploughing) on erosion-prone land can have serious consequences.

- In relation to socio-economic constraints :

An appropriate choice of mechanization must be profitable, taking into account the market value of agricultural products, the high cost of mechanized operations and labor.

Obviously, the low incomes of farmers mean that individuals cannot afford to buy mechanization equipment or pay for mechanized services. Moreover, given the unavailability of land, individual farms are unable to make mechanization profitable.

However, solutions are possible:

a- Implement a rural land plan to facilitate access to land for farmers;

b- Implement a national agricultural credit policy easily accessible to producers. These credits will finance labor and the acquisition of mechanical equipment;

c- Promoting cooperatives for the use of agricultural machinery (CUMA) ;

d- Study beforehand how to introduce motorization, as the direct transition from manual farming to motorization can lead to failure;

e- Train farmers in the rational use of mechanization and farm management.

4- *support provided to farmers extension workers and engineers*

Support can be summed up as capacity-building for farmers. Farmers are trained in :

- organization, operation and management of a village group. Thanks to this training, each village now has at least one well-structured village group;

- Technical itineraries for cotton growing. Thanks to this training, cotton yields are improved;

- Conservation/storage techniques for agricultural produce. This training helped reduce damage to stored produce.

- The use of animal traction. Farmers have and use animal-drawn cultivation.

The difficulties encountered by extension workers and engineers are essentially based on the illiteracy of farmers. For this reason, trainers focus more on practical than theoretical methods. For example, when it comes to technical itineraries, training is carried out on demonstration plots. Ten to fifteen farmers make up a team of participants, to facilitate understanding.

As far as extension workers and engineers are concerned, Benin's agriculture is not very modernized. Producers work hard, but their income is low. The following steps need to be taken:

- mechanize farming to enable producers to better diversify their activities;

- facilitate access to low-interest agricultural credit;

- teach farmers to read and write, to make it easier for them to understand the training courses.

5- *Hypothesis testing*

Hypothesis 1: *There are farming methods adapted to tropical and subtropical regions.*

The intervention of agricultural research and the traditional experience of farmers have made several methods available for crop establishment and farm management in tropical regions. These include :

- knowledge of fertile land ;

- flat ploughing ;

- the practice of line seeding ;

- soil fertilization with mineral and organic fertilizers;

- Crop protection against pests (biological and chemical control) chemical) ;

- Weed control (weeding and herbicide treatment) ;

- The practice of fallowing ;

- The practice of crop rotation.

In practice, these methods remain theoretical and not very operational.

Hypothesis 2: *Methods remain theoretical and not very operational*

Although known, very few farmers actually put into practice farming methods adapted to tropical and subtropical regions.

This is the case, for example, with fallowing, which lasts 2 to 3 years in the villages surveyed, instead of 5 to 10 years.

This is also the case for the yam seed multiplication technique innovated by Beninese agricultural research, which has not yet been implemented by farmers.

Hypothesis 3: *Current production conditions in tropical and subtropical regions are not conducive to the proper application of adapted farming methods.*

In some tropical countries, farmers practise subsistence farming, which is poorly mechanized and dependent on the vagaries of the weather.

Farmers have difficulty accessing specific inputs and agricultural credit (equipment and start-up funds). They also lack information and training. As a result, traditional farming is more widespread. This situation is not conducive to the proper application of farming methods.

CONCLUSION AND RECOMMENDATIONS

Today, agriculture remains an immense reservoir of nutrients for the entire planet.

In most tropical regions, it forms the basis of the economy. It is practised by the majority of the population living in the tropics, generally with rudimentary tools. Much of this agriculture remains traditional because of the use of these tools.

Expected yields and production are often not achieved due to inadequate farming practices, poor climate control and poor soil.

The main difficulties facing tropical agriculture today are the poor practice of certain farming methods adapted to this (tropical) region, due to a lack of training, information and appropriate equipment, and the difficulty of selling produce due to the lack of an appropriate market. To improve these methods and the marketing of products, the following recommendations are made:

- *encourage farmers to use organic fertilizers;*

- *train farmers in the technical itineraries specific to each crop;*

- *facilitate access to agricultural equipment adapted to tropical regions (inputs, work tools);*

- *facilitate access to low-interest agricultural credit;*

- *promote agricultural sectors.*

❖ BIBLIOGRAPHY

* Dominique SOLTNER, (2005), the basics of plant production: General plant science - Volume I **soil and its improvement,** 24th^{me} Edition

* Bureau des études pédagogiques de l'Educatel :

* C.I.E.D. S.A. - Rouen, (2001), **Agriculture Tropicale Tome 1**

* C.I.E.D. S.A. - Rouen, (1999), **Pedology, soil properties**

*	C.I.E.D.		S.A. - Rouen, (2001),__Agriculture Tropicale Economie__

*	C.I.E.D.		S.A. - Rouen, (2001),__Fertilization and Fertilizers__

*	C.I.E.D.		S.A. - Rouen, (2000), __Machinisme agricole Tome1__

*	C.I.E.D.		S.A. - Rouen, (1999), __Tropical Hygiene__

*	J.P.GIROD-Ingénieur CNEARC, (200), __Applied Pedology__

*	D. HOMERIN, (2001), __Agricultural Machinery II__

*	Fred PRINS and Olivier VIGAN and Chrsta KNOBLOCH, (1990), __Contribution à l'étude des stratégies de production dans le système agricole de la zone d'AGOU au TOGO.__

*	Mémento de l'agronome 1996 edition

■	__Soil and plant production__

■	__Improving soil physical properties__

■	__Animal traction and motorization__

■	__Plant breeding and germplasm production__

■	__Fertility management__

■	__Weed control__

* Institut National des Recherches Agricoles du Bénin, (1995), __les sols et leurs potentialités agricoles, les essences forestières,__ Edition1995

*	Hil Kuypers and Anne Mollema and Egger Topper, (1997), __La protection des sols contre l'érosion dans les tropiques (Soil protection against erosion in the tropics).__

*	Pieter Brandjes, Peter Van Dongen and Anneke van der veer, (1995), __Green manure and other forms of soil improvement in tropical countries.__

*	S. HENIN. R. GRAS. G. MONNIER, (1969), __Le profil cultural: l'état__

physique du sol et ses conséquences agronomiques, Second Edition.

* R. DIEHL, (1975), **Agriculture générale, Deuxième** Edition J.- B. BAILLIERE

* J. ERNOULT, (1984), **Agriculture et le petit élevage en zone tropicale,** Edition Saint-Paul

* **Aminata Niane Badiané, Mamadou Khouma and Modou Sène, (2000), Gestion et transformation de la matière organique,** (synthèse des travaux de recherches menés au Sénégal depuis 1945), Edition ISRA.

* Service de protection des végétaux, (1995), **The correct use of pesticides**

* AUTRIQUE and D.PERREAUX, (1989), **Maladies et ravageurs des cultures de la région des grands lacs d'Afrique Centrale**, Second Edition.

* Huges Dupier and Jeanine Simbizi, **Ravages at the camps**

* Huges Dupier, **Hygiene for our plants, prevention is better than cure**

APPENDIX

❖ FIELD SURVEY FORM

This data sheet has been designed to help you carry out your field survey.

It consists of a questionnaire specific to the people to be interviewed.

* Farmers' level

- what are the main crops you grow?

- Which crops earn you the most money?

- Have you ever been a victim of hunger? What were the reasons?

- how do you identify fertile soils?

- which crops respond well to your soil? Why or why not?

- Are there any outlets for your products? If so, how do you assess these markets? If not, why not?

- Do you use selected seeds? If yes, what are the sources of supply? If not, why not?

- how do you fertilize your soil?

- Do you use fallow land? If so, for how long? And what are the results? If not, why not?

- Do you practice rotation? If so, how does it work? What are the results? If not, why not?

- What are the various diseases and attacks affecting your crops and stored goods? How do you combat them and what are the results?

- What types of ploughing do you use? What are the advantages of each type?

- What do you think about the mechanization of agriculture? What's the current situation in your area?

- How do you manage your business to make it profitable?

-       What types of workforce do you have in your locality? Which one is used the most? Why or why not?

-       Do you receive support from NGOs and/or the government to promote your activities? Which ones? How do you rate this support?

-       What diseases affect your health? How do you combat them?

*** Extension and engineering level**

-       What kind of support do you provide to farmers?

-       Are these supports suited to the region? Why do you ask?

-       What approach do you use to reach farmers?

-       How do farmers react to your approach?

-       What results did you achieve?

-       What difficulties do you encounter in your work? How do you solve them?

-       How would you rate Benin's farmers?

-       What do you think of agriculture in tropical regions?

"I hereby certify that I have produced the present document by my own means.

Printed by Books on Demand GmbH, Norderstedt / Germany